Eden... a place in th[e]

Thank you so much for coming to see us here at Eden.

Funny to say, but even after two years we still have to pinch ourselves each day as you appear over the horizon to share this extraordinary place. Eden is made special by its many parents, all fired with the ambition to make a difference however they can.

Eden is about transformation – the transforming of a site as sterile and derelict as only men could create, by restoring it to life through working with nature and human ingenuity. It has been a humbling and hugely uplifting experience for all of us.

The next stage is to interpret our newly created 'Living Theatre'; to shine a light on what we can do to gain a better understanding of our world and how we might best live lightly upon it to the benefit of all. We would not be so impertinent as to suggest that we can make massive changes ourselves, but hope instead to promote a tone of voice, an 'attitude', that brings people together in an apolitical way, and – we're not ashamed to say it – encourages us to think that idealistic is not the same thing as naïve.

As I write this we are embarking on the final phase of the built project, constructions as large as those already on view, and wondering where we get the energy from even to contemplate it, let alone feel the buzz of excitement all over again. The answer, of course, is that there is nothing so life-affirming as finding yourself surrounded by fantastic people working at the peak of their powers towards a common goal.

So please celebrate the successes and cringe with us as we expose our feet of clay, certain in the knowledge that you only make mistakes if you are really trying!

TIM SMIT

A Foundation for the Future
Dr Tony Kendle, Director of the Eden Foundation

The original Garden of Eden is a symbol of paradise, but also of mankind's rejection from it. Historically, conservation policies have assumed that the best-quality environments are those untouched by people, and that environmental care meant keeping people out of them. The real situation is more complex.

Humans have caused problems in the world, but there are also places where we have lived in harmony with nature without complete destruction, and sometimes with a beneficial effect. Communities have already begun to make steps to be effective stewards of the world. The Eden Project is here to showcase those steps, and to tell how each one of us is, already, a global citizen.

We're also here to show that environmental awareness is about quality of life at all levels. The 'environment' is shorthand for issues that impact on us in a thousand ways every day, from the food we eat and the clothes we wear to the weather we enjoy or suffer. Understanding our world better, and the part we play in it, is also about having fun, not about living grey, hair-shirt lives. Our gardens and displays are used as a lens to focus in on the amazing worlds that each one represents; how the politics of the world, for example, lie within a cup of sweet tea. Our plants were chosen because they could tell a story, and every story is there; gruesome, awesome, funny, inspirational, telling of amazing science, giving us hope for food, security, clean technologies and improved health.

We have created wonders to see here in Cornwall, but of course they are just echoes of the real world's diversity, and of the lives that people live in these places. How can we do justice to these lives?

We have tried, first, by creating the best setting we could. You will notice our 'two decent greenhouses', a fitting stage for some of the greatest stories ever told, with room to evoke a sense of the grandeur of the landscapes we are talking about. Already you can stand and feel something of the majesty of the rainforest giants. Already you are able to close your eyes and catch the scents of wild Mediterranean or the Spice Islands.

Just stand by the side of this green oasis and think about what lies underneath. Our garden grows in what was once sterile waste, with slopes and gradients that would make a goat

think twice, let alone our horticultural staff. We have undertaken one of the most ambitious soil-creation projects ever seen, manufacturing a range of different soils for different needs, using recycled wastes and of course avoiding peat.

Why did we give ourselves such a challenge, and why in a large crater? We did it partly for the drama appropriate to the stories we want to tell, and partly to be good neighbours and not affect the skyline. Most of all, however, we wanted Eden to be a symbol of what is possible when people put their mind to the challenge of regeneration and restoration. The pit was once used for china clay extraction, an industry that shaped the economy and the landscape. Today Eden stands at the gateway of a region being reshaped for a positive future.

We should spare a thought for the Eden team. Not only do the horticultural team, for example, have to grow plants from every corner of the world under the most challenging conditions, but no one has ever had to fly a greenhouse of this size before. And let's not forget the construction team, who worked through appalling weather to build something unique, and then there are the people who made the funding work, and those who had the faith to fund it ... the list goes on.

The reality is that everyone associated with this Project has had to make a leap in order to see the impossible become possible. During construction almost every week saw a new reason why Eden couldn't happen. Our biggest thanks should be saved for the local community. Their support in the early days gave us confidence in our belief that the 'build', the effort and determination to make it work despite the odds, is as important as the product. Eden remains a Project, always in evolution.

So where next? At the same time as constructing the Eden Project site, the Eden Trust is also building a new organization. We call it the Foundation, because it underpins everything we do. We will not waste time and energy doing what others already do well, so we are working with a wide network of partners to complement their work. Together we want to tackle the big, difficult questions. We want to identify the barriers to a better understanding and a better world, and start to break them down. The Foundation will be the crucible where we explore approaches, make new alliances and dare to experiment with ideas that may have wider value. It will not be easy, it will not all work – but the bits that do will be fabulous.

This is our Project, this is your Project – we hope you will enjoy joining in.

The Eden Trust

If you believe there should be a place…

…that celebrates life and puts champagne in the veins

…is all about education but doesn't feel like school

…to hold conversations that might just go somewhere

…where research isn't white coats in secret but shared exploration to help us all

…that is a sanctuary for all who think the future too precious to leave to the few – because it belongs to us all.

Then welcome.

That's why we built this place and that's where the money goes.

Most of us have an occasional big dream, but cold reality usually sees it shrivel on the vine. Big dreams cost big money after all, and usually need lots of people, so where do you start if you have nothing to begin with? We dare to dream but we organize to deliver. Eden exists because hundreds of people who could have said no decided to say yes and take a leap of faith. Slowly out of the ether evolved a groundbreaking collaboration between public-sector and private-sector supporters.

Once the keys were handed over we were on our own, totally dependent on public enthusiasm and the generosity of our visitors and supporters, who give us our most valuable asset: the independence to pioneer education programmes and to influence behaviour at all levels and to set new standards of best sustainable practice. The Eden Project is owned by the Eden Trust, a registered charity, and all monies raised go to further its charitable objectives. As a conservation charity we qualify for Gift Aid, and the government gives us 28p for every pound donated – this makes a big difference to what we can achieve.

The Project is unique in that it has deliberately set out to operate in the commercial arena. We believe that only by demonstrating the viability of ethical commerce can we effect real change in the global businesses we aspire to influence.

We try to practise what we preach, use over 200 local suppliers, and have put nearly £300 million of additional revenue into the local community during our first two years. Sustainability is the cornerstone of Eden – it is not a mantra but an operating system whose time has come.

Today we are proud to be described as a social enterprise – a public body ethically managed, informed by the rigour of the marketplace but aiming to bring real social, environmental and economic benefits.

Thank you for your support in making this possible.

Gay Coley, Managing Director

Working together

...and Flora International; Future Harvest; Kneehigh Theatre Company; Landlife; Lost Gardens of Heligan; Plantlife; Rio Tinto; The Guardian; The Sensory Trust; Timber Research and Development Association (TRADA); University of Exeter, School of Education; University of Kent at Canterbury; University of Plymouth; University of Reading; Viridor; Waitrose College; CISCO; Duchy College; English Nature; Falmouth College of Arts; Fauna Camborne School of Mines, University of Exeter; Cornwall

Eden's Partners and Supporters

Eden never knowingly does what others do better. That is why we are privileged to have some of the most forward-thinking and effective organizations working with us to bring our 'Living Theatre' to life. We are delighted to work with those who are determined to solve problems apolitically, fostering improvements in lives, livelihoods and the environment.

In the coming years you will see our partners and supporters taking to the Eden stage, sharing their work, their ambitions, their concerns and, we hope, something of the excitement of making a difference. They, like all of us, come from a wide range of disciplines and have a wide range of views. Eden is a place of many voices – including, of course, yours.

What's the Story?

Eden explores human dependence on plants, and in so doing reveals our global interdependence. This in turn leads us to interpret economic, social and environmental impacts on a wider stage, not only out of curiosity and a shared humanity but also because these factors affect us all.

Eden's interpretation strategy amplifies these ideas by treating the pit as a model world, created from scratch.

First, take one giant clay pit, symbolising all that is sterile, and bring it to life on a huge scale in order to focus the mind on the power of nature and its capacity for healing. From here begin to explore our place in nature by colonising the pit with a range of plants to set the scene.

Second, add superb, gravity-defying architecture that reminds us of the talent and potential of our fellow human beings.

Third, become a living experiment in social organisation to see whether we can develop a culture where the values of home and work are interchangeable.

In short, through great architecture, landscape architecture, horticulture, works of art, artifice and artefact, Eden takes people on a journey of discovery. From the building blocks of life and human dependence on plants through to the power humans have to change their environment for the better, we review many of the issues confronting us, while a range of perspectives on how they might be solved is brought to us by our partners and collaborators.

Eden is not a list of answers. We bring you conversations, stories, views, news, ideas and inspiration …it's more than just a walk round the garden!

If nothing else, we hope that after visiting us you leave with the feeling that there are thousands of people like you who share the vision of a positive future for the world.

So, how do we tell the story?

Seek out intriguing stories, discover new ideas and look at the world afresh.

Plants are fascinating and beautiful, but mute. A major part of our mission is to give them a voice that will be heard by the widest possible audience, to illuminate their world and the essential role they play in ours. We ignore this relationship at our peril. As a project, we explore new ways of communicating in order to stimulate the grey cells and, of course, give you a good day out.

Our exhibits are designed as a series of components. Nicknamed 'Eden Still Life' they include plants, landscapes, artistic installations and signage. The latter includes widgits as well as text. Widgits is a symbol language that puts messages across without words (see Sensory Trust on page 38). You will find exhibits in the Biomes (indoors and out), in the buildings, even in the shop and restaurants. Some are big, some small and little clues are everywhere – check out the sugar in the café.

Over half the exhibits in the landscape depict our dependence on natural environments and wild plants. In this framework are set the stories of the crops and plant products that have made our world. The exhibits grow and change as we develop, making the view through our window on the world an ever-changing one.

Plants set the stage but Eden is really about people. In our 'Eden Live' programme, guides, performers, and visiting partners, supporters and artists bring the dynamic story to life. We still have a lot to communicate and discuss right here in the pit. We also reach out to the rest of the world to take the conversations further and bring back the news.

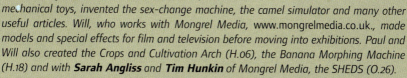

Plant Takeaway Known to the staff as 'dead cat', this exhibit in the Visitor Centre sets the scene demonstrating our dependence on plants. Alan and Enid lose all the plant products in their kitchen … the juice, the fruit, then the table, chairs, books and clothes. The cat loses his milk, and one of his lives! The children's guide-book takes Alan and Enid out across the site to find all their plants again.

Art: Paul Spooner and **Will Jackson**. Paul, who makes mechanical toys, invented the sex-change machine, the camel simulator and many other useful articles. Will, who works with Mongrel Media, www.mongrelmedia.co.uk., made models and special effects for film and television before moving into exhibitions. Paul and Will also created the Crops and Cultivation Arch (H.06), the Banana Morphing Machine (H.18) and with **Sarah Angliss** and **Tim Hunkin** of Mongrel Media, the SHEDS (O.26).

Supporter: NESTA, the National Endowment for Science, Technology and the Arts, enabled us to bring this exhibit to life. www.nesta.org.uk

Eden Still Life

Art

Art is not meant as a substitute for facts and information. The artworks at Eden are signposts to new attitudes and ways of thinking. They are not always comfortable, they are not always beautiful, but they should always be surprising and thought-provoking. We are working with a wide range of artists, locally, nationally and internationally who, along with our in-house team of multi-skilled Designer Makers, breathe life and colour into the Project.

> **Partner: Falmouth College of Arts** is helping with our research into sustainable design products and projects. www.falmouth.ac.uk
>
> **Supporter: South West Arts** have funded research projects to help us build new audiences for art, theatre, music and dance. www.swa.co.uk

Science

Creating Eden's exhibits and growing our plants requires a huge scientific input from a range of disciplines. Scientists and horticulturalists work together to ensure that we display top-class living plant material from which the stories can be told. The team also researches and authenticates Eden's stories and messages.

> **Partners:** We work with many universities and colleges, including the **University of Plymouth**, www.plymouth.ac.uk, and **CSM**, part of the University of Exeter. www.exeter.ac.uk/CSM

Science and Art

Scientists work hand-in-hand with communicators and artists at Eden. Our Science/Art Forum gives them the chance to work together on 'blue-sky' projects. New for 2003: Tim MacMillan's Flybot (a flying robot!) in the Tropics enables us to view the rainforest canopy, and Bill Wroath's beautiful root installations expose the parts of plants that most of us never see.

Angela Easterling, with our science team, is investigating leaf form and function through her beautiful plant photograms.

Eden Live

Our most valuable asset is people. Eden's guides and performers are itching to share their stories, to delight and entertain you.

Eden Themes

Our themed events programme brings together stimulating music, performance, workshops and talks to put champagne in your veins and ideas in your mind. This year we are running 'Metamorphosis' over Easter, 'Earth, Air, Fire, Water' in May and June, 'Look Good, Feel Good' in the summer and 'Global Cultures' in the autumn. Most daily programmes are included in the ticket price. Details of evening ticketed events can be found in the Visitor Centre, on **www.edenproject.com** and in the local press.

Schools and Lifelong Learning Programmes

Our innovative schools education programmes, attracting groups from all over the UK and beyond, link schools into real issues and current global stories. Look out for children chasing clues with the Crazy Chef and clambering through the rain forest on the Don't Forget Your Leech Socks programme. We are also developing programmes and links in Further Education, Higher Education and Lifelong Learning.

Partners and Supporters: The School of Education, Exeter University evaluates our schools education programmes. www.exeter.ac.uk/education
Cornwall College helps to develop staff training and lifelong learning programmes. www.cornrnwall.ac.uk
Creative Partnerships support our educational and creative projects. www.creative-partnerships.com

Beyond Bodelva: taking it home

Look out for *The Eden Trail*, our bestselling children's guidebook, and a growing range of Eden Project books and videos/DVDs. Discover more at **www.edenproject.com**

Friends

Eden Friends now has a membership of 15,000 people from all over the world – active partners in our future. They enjoy unlimited free entry, a quarterly magazine, a programme of special events, talks and workshops and 10% off all retail purchases.

Membership levels are:

Individual	£35.00 p.a.
Joint (2 people living at one address)	£60.00 p.a.
Family (2 adults and up to 3 children in full-time education)	£70.00 p.a.
Lifelong	£1,000.00

For more details please phone 01726 811932, visit the Friends' desk in the Ticketing Hall or email **friendsdesk@edenproject.com**. Join us!

Waste Neutral

The rapidly developing waste and resource debate requires changes in the way we think and act. Eden aims to become Waste Neutral. This is a new, ambitious concept, unique to Eden, adding a new angle to the debate.

Eden is:

1: reducing waste streams. **2:** reusing items wherever possible. **3:** sourcing all remaining items, wherever possible, from materials that can and will be recycled. **4:** adopting a policy of purchasing items that are made from recycled materials, either for use on site or for sale in the shop.

In simple terms, when we buy in a greater weight of products made from recycled materials than the weight of materials we send off to be recycled we will have reached Waste Neutral. This concept can be applied to any organization, community or even individual household.

Do your bit!

To become Waste Neutral we need your help. Check out our Waste Neutral station at Zzub Zzub café, and please use the recycling bins around the site. Please don't use the wrong bin. A major obstacle to recycling is contamination of waste streams – if you throw a sandwich in the plastics or glass bin we have the messy job of getting it out. If you can't find an appropriate bin then please use the general litter bin. We are planning to build a permanent recycling compound, which will include a biodigestion unit for food and organic waste, and wind-row composting.
No waste – no problem.

Pick up a Passport

For just £5 administration fee on top of your admission price your Eden Passport will entitle you to free admission until March 31st, 2004. To get your Passport keep your till receipt and fill in an application form, available from the information desk.

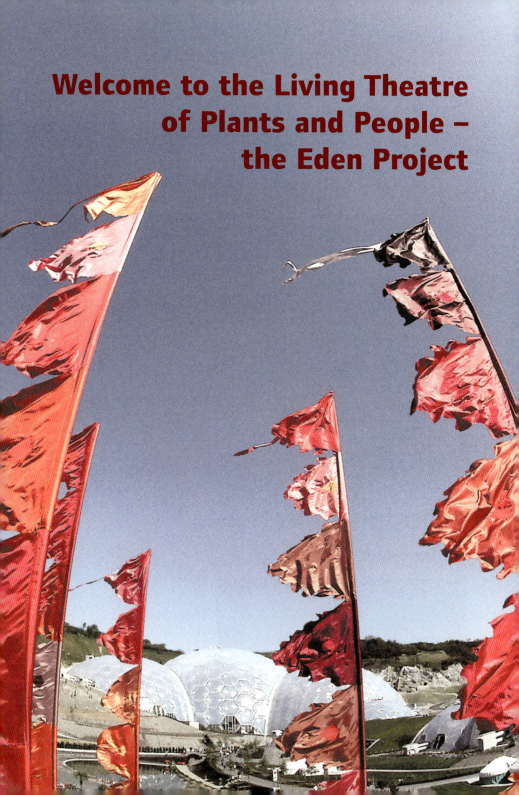

The Outdoor Biome

Please refer to the fold-out map on the inside back cover of this Guide

Here are exhibits and stories about natural landscapes, crops and plants we use every day; plants that nurture us, make history, save lives, entertain us and provide hope, illustrating not only our dependence on plants but our interdependence on each other. Explore Wild Cornwall, the Steppes and the Prairies, vibrant reminders of our place in nature. The pit itself demonstrates regeneration and what people are capable of. Three years ago there were no plants here and no soil; now there are over 3,000 types of plants growing in 85,000 tonnes of soil made from recycled waste.

The landscape will continue to change, as new exhibits appear and others rotate around the site. So we're sorry if things are not always where this Guide says they are, but the maps on site will bring you the latest.

Study the map and take whichever route you like down into the pit. Remember: it's quite a long walk down. A land train runs from the Visitor Centre to the lake and back, and wheelchairs are available on a first-come, first-served basis.

The Zigzag

On this zigzag route down into the pit we are developing some interpretation to put Eden in context, to introduce what it's all about and to get you in the mood. (For now refer to 'What's the Story' on page 6.) We are planning to bring you the story of How the Planet Works in our new Education Centre, when we have built it ... but until then here's a brief introduction.

> **Art: Angus Watt** animates Eden with his sensational flag landscapes, 'My flags celebrate life and are inspired by various plant forms at Eden.' We have a spring, summer, autumn and winter collection.

An Introduction   O.01

Power plants

Plants are living power stations. They turn sunlight into the fuel (carbohydrate) that powers our bodies and they use carbon dioxide and water as the raw materials.

The only by-product: oxygen. The process: photosynthesis (photo = light, synthesize = to combine into a whole).

> **Art.** Sculptor **Kate Munro** has been working with our Green Team to create copper fern forms alongside the zigzag path.
>
> **Art:** Devon sculptor **Paul Anderson** has made tree form sculpture from reclaimed timber and copper, to give our young plantings some height.

And where do the plants get the raw materials? Partly from us. We eat carbohydrate and breathe in oxygen, and produce energy, carbon dioxide and water. It's all a big cycle. Nutrients are cycled round from plants to animals to the soil and back again too. Problems arise when we break these cycles and produce things that can't be recycled. Cue Waste Neutral, conservation, sustainability, Plants for Fuel (O.20) and many more of our exhibits.

Clever plants

Plants can't move (much). If you couldn't move how would you eat, drink, keep warm, keep cool, run away from danger, even reproduce? Plants do all these things. These solar-powered factories also provide the raw materials we need to survive.

Useful plants

Everything we consume has grown or been mined. Plants, and the earth, provide for everyone.

> *From coffee beans to blue jeans, foods to fuels and toys to tools*
> *Pills to potions and soaps to lotions, bridges, homes, cakes and scones*

Plants are even hidden in your car, your camera and your bag; interior panels, photographic film and lipstick are all plant-based. How many have you used today?

Lots of plants … and animals … and people

We are all in this together. Biodiversity – the vast variety of life – makes up an intricate, interdependent web. Living creatures have evolved into many different life forms that live together and share resources. We've named about 1.75 million species, there may be over 10 million. Each one plays a part in the great cycle.

Looking after plants, looking after life

Why conserve? It's a matter of survival: the earth and its plants and animals keep us alive. Research confirms our intuition that the 'wild' places help us feel good.

Eden Live Tent　0.02

Eden is not just about plants. It explores the relationship between plants and people and our interdependence. Eden is a theatre of many voices. Hear the news and views from our guides, performers, partners and supporters in an ever-changing range of workshops, talkshops and displays in our Eden Live Tent.

*Science: **The Scottish Crop Research Institution (SCRI)** sourced the potatoes typical of the Andes for this exhibit.*

Potatoes　0.03

Potatoes, the fourth most important food crop in the world after wheat, rice and maize, are thought to have originated in the Andes. Those from the high mountainous regions are frost-hardy, to survive the freezing night temperatures, and for thousands of years local people have freeze-dried them to make Chuño, dried potato that can be stored. This preservation process gets rid of bitter-tasting 'anti-freeze' compounds.

*Partner: **Future Harvest International Potato Centre, Peru**. Future Harvest bulks up valuable potato types in the high Andes, where there are few insects, and therefore few virus disease problems. These 'clean' potatoes are then reintroduced to the lowland farms to replace diseased stock. www.futureharvest.org*

In Bolivia, Ecuador and Peru, farmers have grown black, orange, dark red, striped, knobbly and smooth potatoes for over 8,000 years. Now attempts are being made to conserve these potatoes and develop products such as crisps and chips, bringing income and economic stability to rural communities.

This year we also feature some of the so-called 'Lost Crops of the Incas', which are grown alongside potatoes in the Andes but are little known elsewhere.

The Making of Garden Flowers　0.04

In the beginning, wild plants were brought to our gardens from across the world. Seeds set and grew and new forms arose. Gardeners sped things up by cross-pollinating (hybridizing) promising parents.

*Art: **Brad Dillon** has created a twisty fence to support sweet peas – in the winter its sinuous shapes will remind us of the summer flowers to come.*

The offspring, not normally found in nature, are called cultivars. Today old forms come back into fashion or have a role in breeding new varieties.

Colours and Dyes   O.05

We started with woad, the true blue, and have just started growing weld and madder to bring you the triumvirate of British dye plants

Indigo *Of blues there is only one real dye – indigo*
William Morris

Indigo, which strengthens fibres and heals skin, has been used by blue Ancient Britons, blue-jeaned teenagers, soldiers, sailors and boys in blue, to name but a few. Indigo is found in several very different plant species across the world. Invisible in the plant, indigo is only formed when the leaves are processed. The dye comes out of the dye-vat yellow and turns blue as it meets the air. Although

Art/science: A team of scientists and artists brought our indigo exhibit to life: Professor Philip John (**University of Reading**) www.reading.ac.uk, Dr Kerry Gilbert and David Cooke (**University of Bristol**) www.bris.ac.uk, of the SPINDIGO project are joined by indigo expert Dr Jenny Balfour Paul (**University of Exeter**) www.exeter.ac.uk, fashion and textile designer **Gary Page** and textile designer and artist **Rebecca Earley**.

natural indigo is still used in many places, synthetic indigo now supplies most of our needs. But new technology means that natural, renewable, non-toxic indigo from European woad plants and subtropical indigo plants, grown elsewhere, can provide work for growers, dyers and designers, and may be able to compete economically with synthetics.

Plants for Taste (Biome Link frontage)   O.06

Vegetables, fruits and herbs can look good, taste good and do you good.

Buy local Vast quantities of fossil fuels are used to transport food thousands of miles round the world. At Eden we reduce food miles and support the local economy by using local produce in our restaurants whenever we can.

Grow your own With a bit of planning you can grow many of your own fruits and vegetables, all year round, whether you live in the country or the city.

Recycle waste To feed the plants that feed us, turn your weeds and organic waste into compost; good for the soil and the plants.

Supporter: The Environment Agency works in many ways – regulating industry, maintaining flood defences and water resources, improving wildlife habitats – and here, helping with our compost display. www.environment-agency.gov.uk

The West Side

Apples   O.07

The Cornish climate and apple trees are not happy bedfellows: the warmth and damp encourage canker, mildew and scab. But over the years horticulturalists have selected seedlings and varieties that tolerate not only disease but acid soil and salt-laden winds. Here we try out Blenheim Orange and Blackamoor Red. No one has ever grown apples down t'pit before!

UK apple growers have declined by a third in the last 6 years. In 1900 around 200 apple varieties were grown in the UK. This figure declined dramatically by the end of the century, but some supermarkets, who sell us 80% of our apples, are now promoting more regional varieties, so things may be on the up. Cornwall has around 80 named varieties and there may be a niche market for products such as apple pies, cider, juice and chutney made from the best of them.

Art: Sue and Pete Hill, brother-and-sister Cornish artists, made our giant Eve out of the soil specially created for Eden, around a timber frame. She has irrigation arteries inside, and her turf skin is of a type used on golf courses..

Plants for Tomorrow's Industries   O.08

We are living at possibly the most exciting time since we came down from the trees. People of all political persuasions agree that, whatever the cause, our wasteful world has to change. We are on the fringe of a revolution and plants lie at the heart of it. Everybody knows that plants make air, food, medicines and so on, but they can also act as green factories providing plant plastics, plant bio-composites and plant oils.

Experiments around the world are focusing on a new green frontier, and now it is within touching distance. Imagine a world engineered by, constructed of, and powered by living things. Could plants take over from fossil fuels?

Supporter: Alternative Crops Technology Interaction Network (ACTIN) works to promote the development of industrial crops. www.actin.co.uk

Art: Pete Hill worked with Eden Designer Makers to create a sculpture from bio-composites and bio-plastics, made mainly from maize plastic and cashew-nut shell oil. Yes, it really is plastic, and it really is made from plants!

Lavender   O.09

Lavender's beautiful scent attracts insects. It has been used to make aromatherapy oils, perfumes, cleaning products, insect repellents, antiseptics and much more. Named from the Latin *lavare*, to wash.

Plants and Pollinators   O.10

Plants can't move (much), so how do they reproduce? Many do it by luring insects, and other animals, to take pollen from one flower to another. The flowers use colour, scent and rewards of nectar and pollen to attract their go-betweens. Over half our food plants worldwide depend on pollinators, so spare a thought for the insects – your lunch may depend on them. Insect–flower relationships are often very specific. This exhibit portrays nature's self-service restaurant: e.g. buddleja for butterflies, moths and bumble bees and fennel for short-tongued flies and beetles.

Art: Robert Bradford, Cornwall-based artist, lecturer and psychotherapist brought us the bee: 'The Bee focuses upon the significance of pollination in the ecology of plant life. Sorry if it looks a bit scary – it must be worse the other way round.'

Plants for Cornish Crops   O.11

Traditionally, Cornwall and the Isles of Scilly are home to early spring crops – bulbs, potatoes, cauliflowers and cut flowers. Times are changing: we can have what we like, whenever we like, flown in from the country round the corner. Some Cornish growers are looking to new markets, such as a diversified range of food crops and innovative industrial crops for oils, dyes and biomass (plants to burn).

Beer and Brewing   O.12

Beer followed in the footsteps of wheat and barley from the Fertile Crescent, the land between the Tigris and the Euphrates (much of modern-day Iraq). In the UK the water was once dirty and therefore dodgy to drink; beer provided a nutritious thirst-quenching alternative. As late as the 1600s many men, women and children in this country drank around 3 litres of weak beer a day.

Art: Reece Ingram, Cornwall-based sculptor, carved the traditional hop poles. Find the ingredients: wheat, barley, yeast and hops, the hop stilt walker, the brewing kettle, the isinglass (the sturgeon's swim bladder) used to clear it – and the magic formula for alcohol...

Tea   0.13

Tea is made from the young leaves of the tea bush, *Camellia sinensis*. Thousands of years ago the Chinese used tea as a medicine. Today its health-giving properties are being extolled again. Tea is grown in around 25 countries in the subtropics and the cool, moist, mountainous tropics, from sea level to 2,100 metres – and here at Eden.

Supporter: Clipper and the Kotada Tea Estate in India bring news and views on fair trade and organic teas. Organic production protects the land, animals and people that live on it, creates a balanced, fertile, environment and a tea free from artificial chemicals. 'Organic production is not a case of old technology, more a case of modern technology moving forward' – Prem Wallia, tea estate director, S. India. www.clipper-teas.com.

The Eden Project supports fair trade as a concept but wishes to point out that it does not imply that all major producers that are not Fair Trade labelled are not trading fairly. We choose to tell the Fair Trade story because it usually involves individual communities that are illustrative of a bigger picture.

Art: Jack Everett, sculptor, designer and builder of low-impact buildings, and **Nicholas de St Croix** designed our tea-leaf house, where you can, on occasion, sup chai and swap tea tales.

Sunflowers   0.14

Now grown on a large scale in Argentina, Russia and the Ukraine, sunflowers are possibly the only major domesticated food crop to originate from temperate North America, where wild sunflowers provided food, fuel and pigment for centuries. Today the seeds provide protein-rich food, oil for cooking, margarine, racing-car engines and paint manufacture, as well as potential for fuel, medicines, cosmetics and plastic. The remaining seed meal feeds livestock, and the husks and stems can even be used to fuel the oil extraction.

Art: Mike Chaikin, Cornish sculptor. 'After looking at time-lapse film of sunflowers tracking the sun by Sarah Darwin (great-great-granddaughter of the man himself) I decided to make a circular field of mechanical copper sunflowers with heads that follow a mechanical sun. The window on the side shows products that are made from sunflowers.'

Eco-engineering   O.15

For centuries we have used plants to bind and heal the earth's fragile skin in areas prone to soil erosion and landslips. Traditional knowledge and trust faded as civil engineering techniques took the stage. Today there is a renewed interest in plants, using heavy construction only where it is essential. At Eden we use this approach.

Hemp   O.16

Canvas sails and ship's ropes
  Tough clothes and bank notes
    Oils and cords, insulation boards
      Soap and Bibles, old masters and fibres

Art: George Fairhurst was a professional tall-ship skipper before becoming a sculptor, so he knows the ropes. To grow hemp at Eden we needed a licence and a physical barrier, so George designed and made the Hemp Fence.

All this from hemp; easy to grow and suited to our climate, hemp needs no chemical input – rare in farming today (in contrast to Cotton, W.16). Hemp products can often take the place of those made from timber or petroleum. So why aren't the fields full of it? Until recently it has had a problem with licensing, even though industrial hemp contains very little of the drug found in marijuana. We look at some of hemp's uses, explore why it fell out of favour, and discover how breeding is bringing it back.

Wheat   O.17

Wheat feeds around a third of the world.

Ancient ancestors These include emmer wheat and goat grass. A chance cross-pollination, by the Caspian Sea, eventually resulted in bread wheat. These ancient ancestors may have a role in breeding disease resistance into future varieties.

The Green Revolution Since the 'Green Revolution', the breeding programme of the 1960s and 70s, wheat yields have increased around 15-fold. Scientists bred short plants that converted the sun's energy into grain rather than stem. These plants produced higher yields and prevented the mass starvation in the developing world predicted before the 1960s. The downside was that these crops required higher inputs from chemical fertilizers and irrigation.

Partner: Future Harvest, an educational charity, supports international agricultural research with farmers in developing countries for a world with less poverty, healthier well-nourished people and a better environment. www.futureharvest.org

Future Harvest Future Harvest International Maize and Wheat Improvement Centre helps to produce disease-resistant, high-yielding wheat varieties for a wide range of global, environmental and cultural conditions. These new varieties can use resources more efficiently and are adapted to low-input cultivation.

Each year drought strikes more than half the area sown to wheat in the developing world. 'Low-till' (ie no-dig) farming leaves the soil and ground cover undisturbed during planting. This saves water (up to 50%), increases harvests, reduces fuel needs (no dig = no tractor) and saves money (fewer weeds = fewer herbicides).

World population is still gradually rising. Can breeding rise to the challenge yet again?

Plants for Rope and Fibre 0.18

For centuries people have made cordage, cloth and ropes from strong plant fibres. Without rope we might never have erected the pyramids or Stonehenge, let alone tied up the runner beans. At Eden we are growing plants for rope: sisal, flax and New Zealand flax.

Artist: George Fairhurst. 'Rope functions when it is used in extension: you pull, and your pull goes all the way down the line to the object. Rope turns nasty rocks or angular boats into soft, round things you can grip and pull.' His huge metal giant is a vertical derrick, holding multiple lines forming a massive rope in his grasp.

Steppe and Prairie 0.19

The prairies, with waving tall grasses and colourful flowers, including Echinacea and Golden Rod, once covered over 1,000,000 square miles (2.5m sq. km.) of midwestern North America. They were created by controlled burning – hence the burnt timbers shown in our landscape – to attract game and ease travelling. Back on the range work is underway to restore and re-create them and let the buffalo roam once more.

Science: University of Sheffield. This exhibit was designed and planted by Dr James Hitchmough, researcher in Landscape Ecology, who studies North American Prairie plants (amongst others) www.shef.ac.uk/landscape

Eastern European and Central Asian steppe grasslands contain spring-to-summer flowering, drought-tolerant grasses and herbaceous perennials. Natural grassland steppe occurs where there is enough moisture for continuous ground cover but not enough for trees to grow, whereas secondary steppe is created by people. Here trees and shrubs might have marched back in without controlled burning and grazing.

Many of the wild ancestors of the Founder Crops in the Fertile Crescent such as wild wheat, barley, lentils and chickpeas originated from the moister fringes of woodland steppe.

Plants for fuel   O.20

When plants are burnt for fuel they release the CO_2 they absorbed during photosynthesis (see O.01 on pages 12-13). Problems arise when we burn 'fossil' fuels – coal and petroleum. The carbon they have stored underground for millions of years once again becomes CO_2, increasing levels in the atmosphere, contributing to climate change.

Renewable plant fuels such as wood (used for fuel worldwide), energy crops grown for fuel (willow coppice and Miscanthus), agricultural wastes (straw and sugar cane waste) and biofuels (bioethanol from sugar cane and biodiesel from rapeseed) are all carbon neutral. If you replant what you harvest the CO_2 levels in the air will remain about the same.

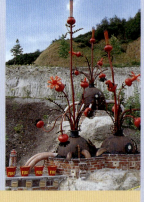

Art: David Kemp, Cornwall-based artist: 'These industrial plants extract energy stored in fossilized plants from ancient forests. They convert raw materials into an astonishing variety of useful products and harmful emissions. Widespread overplanting of industrial plants causes environmental damage and climatic changes.'

But how much land would we need to supply everyone? Do we get more energy out than we put in, when transport and processing equipment are taken into account? Other choices include harnessing wind and water, the hot rocks beneath us, or copying the green plant and going for solar power.

Partner and Supporter: Duchy College and **South West Industrial Crops** research the use of Miscanthus grass as biomass fuel; it fires the college's biomass boiler. www.cornwall.ac.uk/duchy   www.swic.org.uk

Plants in Myth and Folklore   O.21

Myth, folklore, stories and poems keep plants alive in your memory. If remembered and revered they have a far better chance of staying alive in the ground too. Eden performers talk their wares in Myth and Folklore, but also tell stories from around the world right across the site. Everywhere, everyone has their own green man. Pop into the pagoda to catch a tale or two.

*Art: **Pete Hill** and **Kate Munro** have created a magical Story Pavilion, a place to provoke the imagination, to tell old tales and make new ones. Our maze is the same as the ones carved into the granite near Tintagel, and laid out in stones on St Agnes, Isles of Scilly.*

Wild Cornwall   O.22

Cornwall has many habitats, but from heathland to hedge have almost all been shaped by people.

Atlantic woodland

The south-west coast Atlantic woodlands are among the least disturbed of our semi-natural habitats. They contain native oak, willow, ash and hazel, pruned by the wind, dwarfed by the poor soil, and clothed in ferns, mosses and lichens thanks to the clean air. There is a rich diversity in these temperate rainforests, but many species in them are small, green and easily overlooked. English Nature are working to recreate 500 to 1,000 hectares of woodland landscape in the clay area, restoring a habitat that would have been there until the last 100 years or so.

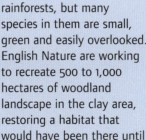

*Art: **Kate Munro** has worked with our Green Team to create the wind-pruned trees, so typical of Cornish uplands. The real plants will take a little longer!*

*Artist: **Chris Drury**, international land artist, has created a Cloud Chamber. Whatever is passing in the sky above will be projected in here – beneath your feet. Magic.*

Cornish heathland

Lowland heath is rarer than rainforest. We created the heathland by farming. It is not artificial, however, but made up of wild plants that thrive under this management system. We lose 15% of it every ten years, mostly due to non-management, but this is changing. Once the heathlands provided sustenance: animal fodder and fuel. Today they provide recreation: fresh air for our lungs, and now sometimes free-range meat and heather beer.

Partner: English Nature's Cornwall Heathland Project – Oberenn Rosvro Kernow – a partnership project between IMERYS, Goonvean and Cornwall County Council, is re-creating over 750 ha. of heathland in mid-Cornwall. The project is part of Tomorrow's Heathland Heritage, a UK-wide programme supported by English Nature and the Heritage Lottery Fund. By 2007 they will have restored or re-created lowland heathland over an area the size of the Isle of Wight. www.english-nature.org.uk

Farmland habitats

Three-quarters of the land in Cornwall is farmed. Responsible farming practices provide rich habitats. Look twice at unsprayed field margins: arable weeds or beautiful wild flowers? Cornish hedges, mini-mountains of soil and rock, shady, sunny, dry and moist, provide a home for hundreds of plants and animals. Indeed some field hedges in Penwith, west Cornwall, are thought to be amongst the oldest in the world, dating to the early Neolithic, 4,000 B.C.

Granite makes fist from hedges
Bracken as brittle as weasel's laugh
Lean-to hawthorn
Fern in constant curtsey
Lichen gloves rock
Heart stopping
Delicious melancholy
Where the land meets the sea
Bird quivers over prey
Moss explodes on rock.

Annamaria Murphy

Partner: Plantlife, a charity dedicated to conserving all forms of plant life in its natural habitat. Plantlife's 'Back from the Brink' programme aims to reverse the declines suffered by threatened wild plants. Their 'Making it Count for People and Plants' programme asks for your help to survey common wild flowers, in your own region, to provide an annual check and help to conserve them and their habitats. www.plantlife.org.uk

Art: Peter Martin and **Sarah Stewart-Smith**, Cornish stone sculptors: 'We have highlighted those species which are most at risk in Cornwall, using local carving stones, serpentine and slate where possible.'

Art: Annamaria Murphy writes for herself and Kneehigh Theatre, one of Eden's partners. More in Tea (O.13) and Poems from Eden (Eden Project Books/Transworld).

Endangered species

The early gentian, *Gentianella anglica*, left, is rare and beautiful. The lichen *Heterodermia isidiophora* is also pretty rare but not very pretty. English Nature, Plantlife and others are working to restore and re-create habitats for these endangered plants. The Cornwall Biodiversity Initiative is a programme that sets priorities and reminds us of our local responsibilities.

The Eden Village

Plants for Fodder   O.23

Fodder crops feed farm animals. So all meat is grass. Well, not quite: fodder crops include herbs and legumes too. Animals get essential minerals from herbs, and legumes make plant food from nitrogen in the air. Pick up the story in the Pulses exhibit (W.12). Fodder legumes revolutionized agriculture by increasing soil fertility, animal stocking rates and farmers' income.

Art: Anthony Frost, Cornish abstract artist, paints a living picture.

Supporters: Raleigh International Millennium Awards helped in the construction of this exhibit. www.raleigh.org.uk

Over the past 30 years farmers turned to intensive production based on artificial nitrogen fertilizers, but today there is renewed interest in fodder crop systems.

Eden Live Tent   O.24

Showing the film of the story behind the project (see page 58). Occasionally a venue for workshops and talkshops.

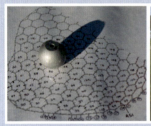

Education Centre   O.25

This hub for our schools in term time is open to the public at weekends and in the holidays for workshops, demonstrations, talks and temporary exhibitions. Details in the Visitor Centre, web and press. This is the temporary home for our successful education programme – we are currently planning and fund-raising for a beautiful and environmentally friendly building to blow everybody's socks off.

SHEDS
(Simulated Housing for Enhanced Displays)   O.26

The 'Story of Rubber', 'Extreme Veg' and many other strange tales can be found in our allotment sheds.

Art: Mongrel Media et al. The SHEDs were conceived, made and written by Mongrel Media: **Sarah Angliss**, **Tim Hunkin**, **Will Jackson** and semi-detached member **Paul Spooner**. Sarah, engineer, composer and author, specializes in curious science. Tim Hunkin, co-founder, is an engineer, Observer cartoonist, TV writer, presenter and exhibition designer. For more details on Paul and Will see Plant Takeaway on page 7.

Coming soon: Plants for Health

Our proposed plants and health exhibit aims to go beyond the historical interpretation of plant medicines. We hope to look at:

Health and nutrition Globe artichokes may stimulate the digestion.

Wild harvest and field-grown plants for medicine Citronella and tea tree may be the new 'natural' health products, but what impact does this have on their wild populations and habitats when demand increases? Field-grown *Digitalis* (foxglove), left, provides digoxin and digitoxin, both used as heart medicines.

Healing landscapes Our ability to regenerate landscapes, such as in this pit, and the ability of natural landscapes to 'heal' us.

Health Futures Eden looks to the future and aims to bring you dynamic changing exhibitions related to current issues in plants and health, for example the controversy around GM vaccines engineered for production in plants, patenting and intellectual property.

Coming soon: Play Structures

We have researched, planned and commissioned some innovative play structures to keep the young and young-at-heart entertained. Play can reconnect you with the natural world and is essential not only for physical health but also social development.

Coming soon: Promise Tunnel

A place to tie up your pledge, your thoughts and your inspiration for the future. Remember your 4 Rs: Reduce, Reuse, Repair, Recycle. Simple things like this can take the pressure off the plants you use every day.

Introduction to the Humid Tropics	H.01
Tropical Islands	H.02
Malaysia	H.03
West Africa	H.04
Tropical South America	H.05
Introduction to Crops and Cultivation	H.06
Cola	H.07
Chewing Gum	H.08
Rubber	H.09
Tropical Timber	H.10
Cocoa and Chocolate	H.11
Palms	H.12
Rice	H.13
Coffee	H.14
Tropical Vegetables	H.15
Sugar	H.16
Mangoes	H.17
Banana	H.18
Tropical Fruit	H.19
Bamboo	H.20
Pineapples	H.21
Pharmaceuticals from the Forest	H.22
Spices	H.23
Cashew	H.24
Tropical Dyes	H.25
Floral Beauty	H.26

Supporter: *We are working with **WaterAid** to bring you the story of water for life.*
www.wateraid.org.uk

The Humid Tropics Biome

Introduction to the Humid Tropics   H.01

This Biome, the largest conservatory in the world, is 240 m long, 110 m wide and 50 m high. It contains over 1,000 plant species, which have shot up since they were planted (from September 2000) and we've lost less than 5%. Misters and waterfalls keep the air moist and ground-level irrigation keeps the soil moist so you won't have to put up with the rainforest's 1,500 mm (60 inches) of rain a year! The air is kept between 18° and 35°C. It gets hotter as you go up, so remove layers as you go.

Supporter: Rainforest Concern protect threatened native forest with exceptional biodiversity. www.rainforestconcern.org

Art: David Kemp built the Tropics Trader at Penzance Dry Dock to provide an informal entrance to the Biome: a reminder that the tropics arrive on our shores every day.

There is an exit point at the Malaysian House (H.03), a cool room in West Africa (H.04) for emergencies, and plenty of seats and water fountains to keep you comfortable on your way.

So join us on a trip through the rainforests of the Tropical Islands, Malaysia, West Africa and tropical South America and discover how local and global people rely on the plants and crops from the tropics. We look at how people are managing the land to meet their needs and conserve the environment. The Eden team visited the regions represented to get advice from our partner communities as to how they would like to be represented.

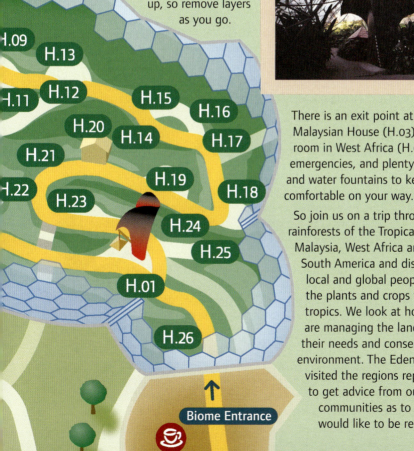

About rainforests (the 'short' guide)

It's a real jungle out there … trees race up to the light, some growing several metres a year, climbers hitch a ride, and epiphytes – orchids, bromeliads (spiky rosettes), aroids and ferns – sunbathe high in the living skyscrapers.

Different species in different rainforests often look similar. Trees have stilts or buttress-roots for support in thin soil, leaves are large and shiny with gutters and 'drip' tips to shed excess water, and on the forest floor many leaves have purple undersides, which act like reflecting mirrors, to make the most of the 2 per cent of light that reaches them. They bounce it back up to get a double dose. Many of these plants now live in our homes, propagated as house plants.

Rain makes rainforests and rainforests make rain: In the tropical rainforests water travels up inside the trees, makes clouds, rains back down, and is taken back up by the trees. It can cycle through seven times before it reaches the sea.

Store cupboard: The rainforests are one of the largest stores of carbon on earth, gathered from the carbon dioxide they take in from the air as they grow.

Weather machine: Changes in areas of rainforest bring climate change; whether it gets hotter, colder, dryer or wetter is still under debate – but we do know that local changes have global consequences.

Forever green? Rainforests are cleared for agriculture, mining, development and timber. But they can also re-grow or be replanted and managed sustainably for the future. Can we win the forests back?

Tropical Islands: Conserving the Land   H.02

Isolated, distant islands are home to some unusual plants and animals. Some are relics of a lost world, extinct everywhere else; others evolve into strange forms, such as giants. All are irreplaceable. Island communities are isolated too, with few resources to support their global responsibility for biodiversity conservation.

Science/Supporter: Seychelles Ministry of the Environment, University of Reading, Darwin Initiative.
Frauke Fleischer-Dogley of the Seychelles Government is a collaborator in the Darwin Initiative Project held by Eden to help save their rare and endangered plants.
moe@seychelles.net www.nbu.ac.uk/darwin

Supporter: St Helena National Trust
Dr Rebecca Cairns-Wicks is pioneering community and non-government contributions to the conservation of local highly endangered species on this small island. www.sthnattrust.org

Climate changes and invasion of aggressive species pose serious threats. Human settlement has devastated some islands, but many countries now have conservation programmes that offer hope.

Protected rare plants in our display include the St Helena Ebony, believed extinct for over a century before two plants were rediscovered clinging to a cliff, and the extraordinary Coco-de-Mer from the Seychelles, the largest seed in the world. We have planted it but it may take years to show growth above ground. However, you can see the seed in our peepshow exhibit. Eden is working with both countries to support their conservation efforts. We also grow common but still remarkable plants, such as the ever-useful coconut and mangrove. Mangrove forests link the land and the sea, protect the coast, provide fuel, timber and fish.

Malaysia:
Orang dan Kebun (People and Garden)   H.03

The home garden around the Malaysian House has fruit trees, herbs, flowers and vegetable beds to provide food security all year round. The garden is zoned – herbs and flowers nearest the house, vegetables in the main garden, fruit and other useful trees further out, and on the other side of the path, the rice paddy. Crops which need tending and picking most regularly are found nearest the house. There are parallels with our own gardens. Winged beans take the place of runner beans; both help to fertilize the soil. Pak choi, taro and rice replace cabbage, carrots and potatoes. In this garden there is even a miracle tree, *Moringa oleifera*, with edible leaves, beans, flowers and roots. It's sometimes called the horseradish tree due to the smell of its roots. The garden also provides building materials, medicines and produce to barter or sell at local markets. It's a self-sufficient backyard larder where local people have selected, collected and bred the best from their superstore – the surrounding rainforest.

Supporters: Infapro, Danum Valley Project, Royal Society. *This garden mainly reflects research on five smallholdings at Kampong Tampinau, a small village in Sabah, Malaysian Borneo.* www.ssl.sabah.gov.my/project/danum.htm www.royalsoc.ac.uk

Art: Mark Biddle, **Hamish Thomson**, **Brian Lloyd** and the **Eden Project team** designed and built the Malaysian House using timber, rattan and bamboo. Mark was a sustainable land use consultant in S.E. Asia before joining Eden. Hamish, a builder, farmer and craftsman like Mark, also has a background in permaculture. Brian, from Jungle Giants, has spent many years working with bamboo.

Science: Simon Platten, Social and Environmental Anthropologist, **University of Kent at Canterbury**, is involved with acquiring and identifying Eden's economic botany artefacts displayed throughout the site. www.ukc.ac.uk

West Africa: Managing the Land   H.04

How do you feed the soil, feed yourselves and replant the forest simultaneously?

Alley cropping: Maize, sorghum and other crops are planted between trees. The trees alleviate soil erosion (especially on slopes), increase soil fertility (if they are legumes or if branches are used as a mulch) and often produce a crop themselves. Vetiver grass, also used to stabilize the land, creates less shade than trees. Originally from India, it has been used for 3,000 years as a perfume, insect repellent and boundary hedge.

Supporters: This exhibit is based on a visit to, and a research project with, **Ndoumdjom village**, an agroforestry region in the Tikar plain in the Adamaoua Province of Cameroon. Thanks to **REFOCON** (a Cameroon NGO) for their assistance.

Art: The African rondavel. The Eden team, aided both by colleagues in Cameroon and by Cornish archaeologist **Jacqui Wood**, have re-created an African rondavel, which provides temporary accommodation for farmers whilst tending cattle and/or crops. It contains the bare essentials: shelter, a sleeping area, a fireplace and a storage place.

The taungya system: Crops like coffee, that need shade, are grown beneath useful trees such as *Prunus africana* and kapok in a tiered production system. As the trees grow bigger the garden once more becomes forest: in this case a productive crop forest. These agroforestry systems enable local people to produce food crops and cash crops whilst improving their land and environment for the future.

As you leave this area of the Biome the path splits. The high road takes you past the waterfall. The view is excellent but the path can get crowded and there are steep steps to get down at the other end. The low road is flat and takes you past the shifting cultivation exhibit in tropical South America. The paths meet by the lily pond.

Tropical South America: Shifting Cultivation and Plant-Gathering   H.05

In the rainforest, villagers practise shifting cultivation, moving their garden on to a new plot every year, rotating round the village on a 14-year cycle. They cut about an acre of forest, let it dry and then burn it to provide fertility from the ashes. Staple crops such as cassava and sweet potatoes are then planted. Cassava often contains prussic acid or hydrocyanide, a poison that has to be washed out before cooking. After harvesting the land is left and is soon reclaimed by the forest. The forest also provides a natural garden. Plants are collected for food, fuel, medicine and materials.

A Peruvian shaman showed us how myths and stories help to keep the knowledge alive in local cultures and the plants alive in their forests.

Supporters: Environmental Protection Agency (EPA), Guyana. *Thanks to the EPA,* **Iwokrama Field Centre** *and the people of* **Fairview** *who taught us about their forests and gardens.* www.sdnp.org.gy/epa

Science: University of the West of England, Bristol. *Dr Mark Johnston, senior lecturer studying the utilization and conservation of tropical forests, is assisting with our exhibit.* www.uwe.ac.uk/fas

Art: Francisco Montes Shuna *and* **Yolanda Panduro Baneo**, *inspirational shamanic artists from Peru, painted the cliff face in the Biome with pictures of the spirit lives of rainforest plants. Their visit was made possible by the* **October Gallery**, *London, and* **Visiting Arts**. www.october.gallery.ukgateway.net www.britishcouncil.org/visitingarts

Introduction to Crops and Cultivation   H.06

Madame Wealth
Every day rainforests touch our lives, and we touch them. Rubber gloves, oils in processed foods, chocolate, lipstick gloss and food colouring can all come from the tropics. Many of the things that we do and buy affect the future of the rainforest.

Art: **Paul Spooner** and **Will Jackson**, *Cornish automata makers, crafted the Crops and Cultivation arch, where cola pods, cocoa pods and chicle sap meet cola drinks, cupcakes and chewing gum.*

Cola   H.07

Probably the best-known Latin name in the world. *Cola nitida*, an African tree with caffeine-rich seeds, is part of the age-old culture of West Africa. Cola, a sparkling flavoured drink, is part of a new global culture.

Chewing Gum   H.08

A milky latex, chicle, harvested from the sapodilla tree, *Manilkara zapota*, can be made into chewing gum. This doesn't harm the tree (if it is not overtapped) or the forest, provides a living for local people and makes gum that can clean your teeth. The trees also yield tasty fruit.

Rubber   H.09

Elastic, waterproof and mouldable – from work to play and transport to health, rubber has played its part. The milky latex of the rubber tree, *Hevea brasiliensis*, has been tapped for centuries in tropical South America to make rubber boots, bottles and balls. In the 18th century European scientists stepped in with waterproof clothing and catheters. Then came the car and with our new passion for travel a spiralling demand for rubber to keep those tyres turning. Thousands of indigenous people were killed in the greedy race to tap from the wild. A frantic global search for plants producing alternatives was followed by the rise of a new industry in Asia: cultivated rubber. More demand, more supply, oversupply and complex schemes to restrict output caused rivalry between players. Wars restricted supplies and introduced synthetic rubber. Rises in petrol prices, AIDS and a need to find a natural solution: condoms and rubber gloves. Natural rubber bounced back. Where next?

Tropical Timber   H.10

Tropical forests provide us with a range of products, the most important being timber. The Industrial Revolution led to widespread commercial exploitation. Now the threat to the world's forest resources has been recognized, consumers and companies, especially from Europe and N. America, seek tropical timber from sustainably managed forests. Improved harvesting methods or reduced-impact logging (RIL) help to make a credible assessment of how much can be cut out without harming the forest.

Wood can be an environmentally friendly, renewable resource. One way to help is to buy from a certified source. Several independent organisations have certification schemes and systems to assess forest management practice. Ideally, certified timber can be tracked from the raw material to the end product. The Forest Stewardship Council runs the largest, globally recognized certification system. Over 31 million hectares of forest in 56 countries are FSC certified, but there is still a way to go especially in tropical regions.

Partners and supportrs: TRADA, NRI, NR Int. Ltd, FSC. The Timber Research and Development Association (TRADA) is an internationally recognized centre of excellence on the specification and use of timber in wood products. The Natural Resources Institute (NRI) carries out research, development and training to promote efficient management and use of renewable natural resources in support of sustainable livelihoods. Natural Resources International Ltd works in collaboration with governments and development agencies in the management of natural resources development programmes. The FSC logo identifies timber and wood products from independently certified, well-managed forests. www.trada.co.uk www.nri.org www.nrinternational.com www.fsc-uk.org

Cocoa and Chocolate   H.11

Cocoa beans, brewed up with chillies, were drunk by Mayan and Aztec nobles. The Latin name, *Theobroma cacao*, means food or drink of the gods. Cocoa trees started life in South America's rainforests, but today most comes from smallholdings in West Africa. In 1999 the global chocolate industry decided to promote an international initiative to create a worldwide programme of research and development on sustainable cocoa production. Rather than clear new areas of natural vegetation the aim is to replant existing cocoa-growing areas with improved varieties that are more productive and disease-resistant. British chocolate manufacturers, through their trade association BCCCA, are involved in a programme to collect and conserve wild cocoa varieties.

Research is also looking at production systems in forest gardens and smallholdings that benefit the environment and the producer. Fifteen million farmers throughout the tropics will grow more reliable, better-quality cocoa, and we can still have our chocolate – looking after the tropical rainforests and the plants they contain can help to provide livelihoods *and* luxuries.

Supporters: Green and Black's and the **Biscuit, Cake, Chocolate and Confectionery Alliance**. We are working with Green and Black's to bring you the story of their cocoa production from Belize, and with the BCCCA to tell the story of the global research and development programme on better cocoa production. www.greenandblacks.com www.bccca.org.uk

Palms H.12

The stems, leaves, trunks, sap and fruits of many palms provide walls and thatch, ropes and boats, palm hearts and dates, coconuts and sago, sugar and wine, cooking oil and much more, making a huge contribution to the livelihoods of local people. Coconuts also 'wash up' on European shores, used mainly for their flesh (copra) and fibre (coir). We may be more familiar with coconut chocolate bars, piña coladas, hair conditioners, potting composts and doormats. However, on the international market the oil palm reigns supreme. It produces palm oil, found in our processed foods, cleaning products and cosmetics. Supply chases demand and plantations march into the rainforest. Plantation work is hard and dirty, but people work dirty and hard to get the world on satellite TV. 'Don't cut down the forest,' we say. Who are we to talk? We already have our TV dream.

Partner: Natural Resources Institute
See also Tropical Timber (H.10).

Where next? New initiatives are slowly emerging; projects to explore planting oil palms on degraded land rather than newly felled virgin rainforest, and new co-operatives to enable workers to control their own destinies.

Rice

Age-old respect

We see a man in the moon. In Vietnam they look at the moon and see the Rice Goddess, stacking her freshly harvested rice in the shade of a Bo tree. Gold, diamonds and pearls are our treasures. But in Chinese tradition, five grains are precious jewels; the first of these is rice. Why? Because this grain has fed more people over a longer period than any other crop, and today nourishes around half the people in the world.

Partner: Future Harvest International Rice Research Institute. In the 1960s population growth led to pressure on food supplies, particularly in the developing world. Financial and scientific research was mobilized to improve crop productivity. Future Harvest's International Rice Research Institute was set up in the Philippines and in the next thirty years the global rice harvest doubled. www.future-harvest.org

The Green Revolution and breeding

In the 'Green Revolution' (see Wheat, O.17) different rice landraces (ancient types of crop plants, whose genetic diversity helps them adapt to their growing environment) were crossed to produce high-yielding semi-dwarf varieties. The first 'miracle rice', IR8, released in 1966, was a cross between a tall Indonesian variety and a semi-dwarf Taiwanese variety. Ganesan, an Indian farmer, was so impressed with his bumper IR8 rice harvest that he named his son after it. The increased income from the crop later enabled Ganesan to send IR8 (IR-ettu in Tamil) to college. In 1990, IR-ettu got his BSc. in Zoology.

Art: Cornish artists **Phil Booth** and **Louise Thorn** worked with Japanese rice-straw sculptors **Eio Okumura** and **Prof. Yoshihisa Fujita** to make our Shimenawa. This giant decorative rice-straw rope was traditionally made after the Japanese rice harvest to celebrate the Shinto gods. The Shinto religion pays great respect to nature, promoting harmony between humans, plants, animals and the landscape. The rice goddess, not typically Japanese, represents the belief systems in other parts of Asia, where rice is personified as female. Here she watches over the many rice landraces, one or more of which may be the parent of the 'rice of the future'.

Where next?

Between now and 2020, around 1.2 billion extra rice consumers may be born in Asia alone. IRRI scientists are looking to some of the hundred thousand rice landraces for assistance in future breeding programmes. They have even drawn up a blueprint for the ideal plant, with high yields, stiffer, fewer stems, increased seed-head size and seed weight.

Coffee   H.14

Coffee is not only big business; it has been a driving force in history. Starting life in Ethiopia, coffee travelled to the Yemen, took a pilgrimage to Mecca, wound up the whirling dervishes and gave birth to the coffee house in the Middle East.

Exchanging news and views, wheeling and dealing, chin-wagging, and even plotting – these cafés provided the place, and the coffee the stimulation. By the 1600s coffee and its houses reached Britain and continued to spawn intellect and commerce. Lloyds of London, the *Tatler* and the Royal Society all started life in coffee houses.

Supporters: Fairtrade Foundation (FTF)**, Cafédirect, CABI Commodities.** *The FTF, who work with coffee growers to ensure they get a better deal, tell their story at Eden. Cafédirect are putting us in touch with coffee growers and processors who are telling the story from their point of view. 'Fairtrade is very important to us because we receive a fair price, which helps us meet the financial needs of our families. Last year this went to help my children with their university education and, for the younger ones, to buy schoolbooks and uniforms. Then if someone in our family is ill we are able to pay for a visit to the doctor and to buy medicines.' Frolian Olivera Gutiérrez, coffee farmer, Peru.* www.fairtrade.org.uk www.cafedirect.co.uk *CABI Commodities promote profitable, healthy and environmentally safe commodity production for resource-poor farmers through information, research and training.* www.CABI-Commodities.org

Partner: The Sensory Trust *has been working with the Designer Makers team to develop inclusive design approaches to interpreting the coffee exhibit.* www.sensorytrust.org.uk

Luxury or necessity?

Coffee fuelled the industrial age, and today is probably the most valuable tropical product on the world market, amounting to an estimated US$70 billion of retail sales per year. At the beginning of the chain, things are often less rosy. Many beans are still picked by hand, labour is high and income low.

Tropical Vegetables   H.15

Tubers, stems, leaves and flowers: a varied and vital part of the tropical diet providing proteins, starch, vitamins and minerals. Giant tubers of the yam form a third of the tropical African diet. Sweet potato has edible leaves and tubers. The latter now crop up in our supermarkets. Tropical legumes such as Jack bean and Lima bean feed people and the soil (see W.12). Snake gourd fruits, exotic cousins to the courgette, can grow to 120 cm long. Young loofah fruits are very tasty, mature fruits make better back-scratchers.

Sugar   H.16

During the Renaissance, people used around a teaspoonful of sugar a year; now we use over 2,000 times that amount: around 135 million tonnes! Some 60% comes from sugar cane grown in the tropics, most of the rest comes from sugar beet, which grows in the temperate regions.

As well as providing sweet delights, sugar can be made into fuel and plant plastics. Molasses, the liquor remaining after sucrose has been crystallized from the hot, concentrated cane juice, provides animal feed and rum. The fibre too makes fuel, paper and fibreboard. Today the tables are turning: some countries are growing sugar for fibre for fuel, the sugar itself becoming the useful by-product.

Mangoes   H.17

Mango, one of the earliest cultivated fruits, has been grown in India for about 5,000 years. At least 500 varieties are grown there! With its strong, sweet, flavour, it is the second largest tropical fruit crop. Mangoes can be round, oval, or oblong and the fruit colour can vary from green to yellow-orange. When buying a mango, select an unblemished, firm fruit. It will ripen in three to five days at room temperature. Mangoes keep well in the fridge for about a week.

Banana   H.18

Totally tropical taste Over 75% of the world's bananas are consumed on home ground in the tropics. Different types are used as sweet and savoury foods, and for beer, cloth, roofing material and more. In Africa 70 million people eat banana and plantain every day as part of their staple diet.

Whoops, have a banana Around 20% of the world's bananas, mainly Dwarf Cavendish, are exported. When this yellow energy booster first appeared in Europe, some ate them with the skins on. Comedy was used as a marketing tool, which insured its popularity.

Save the banana Bananas can get sick, and disease-resistant varieties need to be introduced. Banana diversity, a vital resource for poor farmers throughout the tropics, needs conserving - but how? Plant collections are often stored in seedbanks and most bananas don't have seeds!

Science fiction or science future?

The Future Harvest Centre, INIBAP, holds the largest genebank of banana tissue in the world. The collection is held 'in trust' for the public good and its accessions are freely available. Recently advances have been made in deep freezing (cryo-preserving) plant material. INIBAP is working to cryopreserve the entire banana collection of over 1,000 accessions.

Partner: *Future Harvest International Network for the Improvement of Banana and Plantain (INIBAP).*

Art: Paul Spooner and **Will Jackson**. *Design your own banana, courtesy of Paul and Will's Banana Morphing Machine. Also see Crops and Cultivation (H.06).*

Tropical Fruit   H.19

Papayas, rambutans, jaboticaba and many 'new' tropical fruits are becoming more popular in supermarkets, but the effects are often felt across the world. We look at the future: what will the impact be of these imports, both here and there?

Bamboo   H.20

This green gold of the east is used by half the world's people. It makes homes and furniture, food and fuel, music and medicine, paper and poles, skyscraping scaffolding and gorge-spanning suspension bridges. Ancient Chinese philosophers praised it. Their writers called it 'the gentleman' and their artists used it for canvas, brush and subject. Bamboo's hollow tubes make it strong but light. Within its tissues short, tough fibres sit in a resilient matrix, providing nature's version of fibreglass.

Grow your own house in five years: fast-growing bamboos are ideal materials for low-cost, low-impact, earthquake-resistant houses. This renewable resource provides materials and employment and unites science and art, rich and poor, high tech and low tech, city and country – a real bridge-builder.

*Art: Housings and Hazards, from Exeter, work to make affordable, hazard-resistant housing available to vulnerable rural communities around the world. They led the construction of our new two-storey bamboo house frame. www.HazardResistantHousing.com This design was originally created by renowned Colombian architect **Simon Velez**.*

Pineapples   H.21

Pineapples are herbaceous perennials from South America. They take 3 to 4 years to fruit. Grow your own by cutting (or screwing) the top off a fruit and planting it in a pot in a warm place. Fruits develop from tiny, lavender flowers that grow from the centre of the leaves. Now a familiar sight, this fruit was highly prized at Victorian dinner parties. They are still grown in Victorian style at the Lost Gardens of Heligan, near Eden.

Many pineapples are produced on large farms hundreds of hectares in size. Crops grown on the same plot each year may use large amounts of fertilizers and pesticides, so scientists are looking to GM technology to reduce the amount of chemicals.

Pineapples produce alcohol, silk (pina fibre from the leaves), candles, animal feed and medicines as well as chunks and rings.

Pharmaceuticals from the Forest   H.22

Medicine chest Only 5% of rainforest plants have been tested for their medicinal value. Conserving the rainforest's medicine chest may provide cures for today's illnesses.

From forest to field Plants such as Madagascan periwinkle, source of the alkaloids used in treating leukaemias and Hodgkin's Lymphoma, are grown as a field crop since the alkaloids they contain are too difficult or expensive to synthesize.

Serendipity Drugs are often discovered by good fortune rather than mass plant screening. The Madagascan periwinkle was used locally as a folk remedy for diabetes. Research into its effectiveness showed it was no use for diabetes but did have cytotoxic activity. The symbol of the Lymphoma Society is a periwinkle!

Spices   H.23

Today spices are cheap. In the past they were worth their weight in gold and shaped the world as we know it.

Spice and death: Nutmeg was thought to cure the bubonic plague. Yet in the mid-1300s it was along the spice route from Central Asia that the Black Death first travelled to Europe. It killed a third of the population in five years.

Secrets and lies: The Arabs monopolized the land-based spice trade to the West until the 15th century. Western Europeans were spun yarns to keep their traders away – of Arabs fishing for spices by moonlight, of cinnamon harvested from the nests of ferocious birds and of boiling seas. In the early days they also thought the world was flat.

From camels to gunboats: In the 16th and 17th centuries the Dutch, Portuguese and English literally blasted each other out of the water in their race to the Spice Islands: the Moluccas of the East Indies. This brutal trade made the fortunes of many places, ports and people, and partially funded the industrialization of Europe.

Disappearing island: In 1667 at the Treaty of Breda, after years of battles deciding who 'owned' what, the English relinquished their claims on Run, a tiny island in the East Indies, and the Dutch theirs on New York (formerly New Amsterdam), a small town on the island of Manhattan. A nutmeg island, no longer on the map, was swapped for the Big Apple.

Art: Bill Mitchell (Artistic Director of Kneehigh Theatre) and **Dave Mynne** (artist/graphic designer, Penzance) made the spice boat, to continue the adventure on the high seas. 'The spice exhibit has got everything: adventure, mystery, early espionage, greed, politics, and passion. It's as much about people as it is about plants.' See if you can solve its riddles.

Cashew   H.24

These fast-growing, drought-tolerant trees produce nuts after 3 years and can live for 50. Seeds, unusually suspended below the fruit, provide the highly prized nuts, and the shells', cashew-nut shell liquor (CNSL). Why are cashews more expensive than peanuts? Because roasting, shelling and cleaning of the kernels is a delicate and laborious process and CNSL is a highly-corrosive substance, exploited traditionally in the treatment of ringworm and warts. Globally, CNSL crops up in marine paints, heatproof enamels, brake pads and more recently as a resin in 'ecocomposites'. Traditionally bio-composites are made from plant fibres, such as hemp, embedded into fossil-fuel-based resins. We have our own CNSL tree in Plants for Tomorrow's Industries (O.08).

Supporter: Technoserve provided our trees. This NGO has been supporting the development and expansion of the cashew industry in Ghana since 1994.

Tropical Dyes   H.25

Under the skin of our cultural diversity we have much in common – plants provide the coloured backdrop to our lives. Bark, leaves, roots and flowers dyeing our skin, food, clothes and hair. Brazil wood was originally a name given to sappan wood, from Asia. Why? Because it means red-hot coals – the colour of the tree's dye. When the Portuguese landed in South America (around 1500) they found other red dye trees and named the land Brazil after the dye wood they already knew.

The seeds from the South American tree *Bixa orellana* (annatto) provide a red dye used locally as a body paint, insect repellent and hair dye. We use this tasteless dye from plantation-grown trees as a food colouring in sweets and cheese.

Floral Beauty   H.26

Beautiful plants such as Aristolochias, Oxalis and Hibiscus astonish with their amazing variety and remind us of exotic holidays. The plants we know from our kitchen windowsills bring the rainforest into our homes.

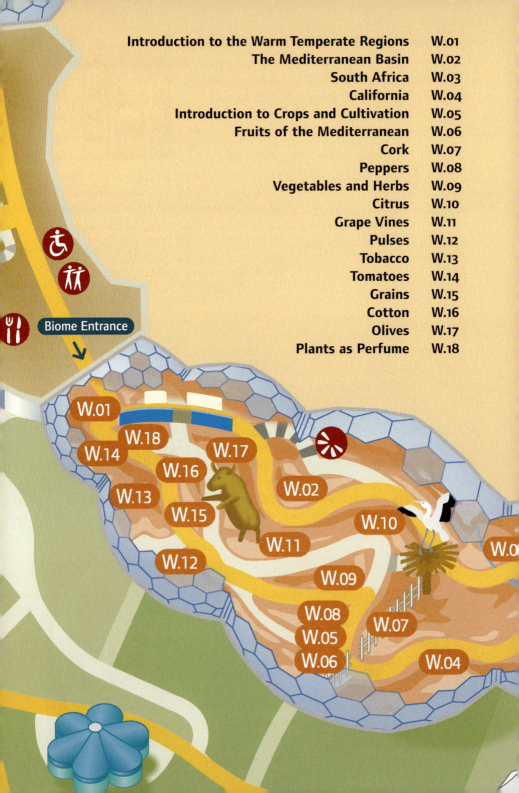

Introduction to the Warm Temperate Regions	W.01
The Mediterranean Basin	W.02
South Africa	W.03
California	W.04
Introduction to Crops and Cultivation	W.05
Fruits of the Mediterranean	W.06
Cork	W.07
Peppers	W.08
Vegetables and Herbs	W.09
Citrus	W.10
Grape Vines	W.11
Pulses	W.12
Tobacco	W.13
Tomatoes	W.14
Grains	W.15
Cotton	W.16
Olives	W.17
Plants as Perfume	W.18

The Warm Temperate Biome

Introduction to the Warm Temperate Regions   W.01

Halfway between our wet green woods and the world's deserts are the warm temperate regions. An important part of these are the Mediterranean regions, characterized by hot dry summers and cool wet winters. They are found on the western sides of continents between 30-40 degrees N or S latitude, and are caused by cold ocean currents, trade winds and topography. They include parts of California and South Africa, S.W. Australia and Chile as well as the Med. itself – the warm, sunny holiday places. Natural gardens bloom in a gardener's nightmare of drought, scorching sun, poor, thin soils and fire. Plants have spines, waxy evergreen leaves or small, grey, hairy leaves, all of which serve to protect. Shrubs are more common than trees, bulbs hide below the summer-scorched soil, and annuals bloom in a riot of colour after winter rains.

In this Biome the air is kept between 15 and 25°C in the summer and a minimum of 10°C in the winter. Breathe in … the scent comes from the plants' protective oils. These may act as bug repellents and vapour barriers to reduce water loss. Our Biomes, unlike glass, transmit UV light, which also increases plant oils – so on sunny days don't forget to protect yourself too.

Conservation and cultivation

The plants are tough but their environments are fragile. Intensive grazing has caused soil erosion, imported plants have threatened native species and land has been developed. We take you on a journey through the 'wild' landscapes of the Med., South Africa and California (many of which have been shaped by mankind through the centuries) and on to the intensively cultivated areas where water, food and shade have created a kitchen garden for the supermarkets of the world. Cornucopia?

The Mediterranean Basin   W.02

When you don't cultivate the land in the Mediterranean the land dies.   Fernand Braudel

Just as you enter the Mediterranean area the path splits. The upper route, with steps, takes you to The Outlook, where you get a good view across the whole Biome. The lower route, following the golden mosaic path, has no steps. The paths meet before you enter South Africa.

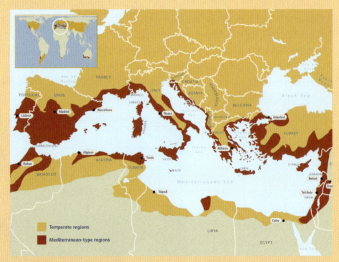

Culture's cradle

Natural Mediterranean vegetation has been cut for timber and firewood and cleared to plant crops for hundreds of thousands of years. One third of Crete is terraced – witness to the long period of human effort to grow food here. Without the olive and the vine, Mediterranean civilization might never have begun. The 'natural' landscape we see today is the product of both nature and mankind.

Classical wisdom

The words of the ancient Greeks still ring true and can still be heeded.

> 'What now remains compared with what then existed is like the skeleton of a sick man, all the fat and soft earth having wasted away, and only the bare framework of the land being left.'   Plato, Greek philosopher, describing the deforestation of Attica, around 400 BC.

> 'The Earth conceives and yields her harvest … but if anything goes wrong, it is not deity we should blame, but humanity, who have not ordered their lives correctly.'   Plato

> 'Earth is a goddess and teaches justice to those who can learn, for the better she is served, the more good things she gives in return.'   Xenophon, Greek general and farmer, 400 BC.

Today's landscapes: Maquis and Garrigue

The French underground movement in World War II was called the Maquis, because they hid out in this habitat with its prickly oak, juniper, wild olive, laurel, myrtle, tree heather and broom. Partially man-made, maquis can grow into woods if not grazed. Garrigue, with shorter plants and more herbs, is found where there is less soil moisture. Both habitats contain unique plants, insects and reptiles, but can get overlooked, having no spectacular birds or mammals. Do we only save the pretty things?

Today's landscapes: traditional olive groves

The ancient terraced olive grove supports far more animal species (insects, reptiles, birds, bats, etc.) than a pine forest. However, people are leaving mountain farms for work on the coast. The goats remain, grazing and knocking down the terraces. What to do? One solution proposes focused subsidies, managed locally, to sustain traditional crops and farming. Others argue that subsidies themselves are the heart of the problem. Buying traditional foods and natural products, seeking out quality and taste, farm holidays: all can help conserve these fragile environments and communities.

Art: Elaine Goodwin, *mosaic artist, created the Liquid Gold pathway as a celebration of the long tradition of olive oil as a symbol of light, life and divinity. Look out for the subtle images of doves – one for each Mediterranean nation.*

South Africa

W.03

The Fynbos: botanical hotspot

The Cape Floral Kingdom has a richer density of different plant species than anywhere else on earth, globally important and very fragile. Fynbos, with around 7,000 plant species (5,000 of which occur nowhere else in the world), covers 80% of this Kingdom, stretching for 46,000 sq. km. either side of the Cape, in a band no more than 200 km. from the southern coast. 'Fain-boss' is Afrikaans for 'fine bush' and refers to the evergreen, fire-prone shrubs that live in this nutrient-poor soil. Formed millions of years ago from the ashes of drought-stressed forests, the Fynbos has been extensively fire-managed for conservation since the 1960s.

> **Partner: Fauna and Flora International** (FFI) bought 550 hectares of fynbos, Flower Valley, in 1999, saving it from being ploughed up to grow vines. They then established sustainable harvesting methods for cut flowers here and on surrounding farms, influencing the management and conservation of fynbos over 25,000 hectares. Profits from cut flower sales are invested in education, the business and research. FFI is the world's oldest international conservation body, whose remit is to protect the entire spectrum of endangered animals and plants worldwide. www.fauna-flora.org

Plant groups include restioids (rush-like plants), proteoids (including feather-like Proteas), heather-like plants and stunning lilies, orchids and irises. The fynbos is threatened by urban spread and development, pollution, agricultural conversion and invasive alien tree species.

Little Karoo

This long, narrow, semi-arid valley, behind the southernmost coastal mountain range of South Africa, lies between the Fynbos and Namaqualand. It bakes to 50°C in the summer, freezes in winter, and droughts are the rule. Today the valley is irrigated for crops, but the surrounding hills house ice plants, aloes and many types of daisy. The Little Karoo was once the natural home of the ostrich, which is now farmed there.

Namaqualand

This red desert, about 250 miles north of Cape Town, is at the dry end of the Mediterranean climate. After rain it blooms into a multi-coloured carpet. The seeds are temperature-specific so different flowers germinate in different years depending at what time the rains come. They take it in turns to share out the scarce water. Many plants in these regions are at risk of extinction, yet you will find their descendants, including daisies, ericas, geraniums, gladioli and lilies, in your garden.

California

W.04

The various Mediterranean climates of California lie between San Francisco and Baja California and include the thin coastal strip, the chaparral, the foothills of the mountain ranges and the extensive grasslands, marshes and now intensively cropped areas of the vast Californian valleys. California once had so rich a natural harvest that the Native American tribes had no need to develop agriculture and a third of all the Indians in North America lived there. Today valleys of central California are the kitchen gardens of America and some of the most intensively farmed regions in the world.

Cowboys and Indians

California is the birthplace of blue Ceanothus and Californian poppies that we now see in our own back gardens. Out in the dense, tough, spiky chaparral live less familiar faces – scrub oak, buck bush, toyon, white sage and many more. The scrub oak, alias the 'chaparro', gave its name to chaparreros or 'chaps', worn by cowboys to protect their legs when riding through on horseback. Chaparral, grassland and the open oak forests were the results of thousands of years of controlled burning by Native Americans. Today battles are not between cowboys and Indians, but eco-

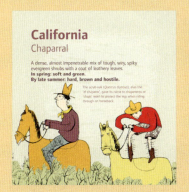

logists, conservationists, developers and farmers. Water for irrigation is so valuable that environmental activists have taken out court orders against farmers to make them leave minimum flows in rivers (see Citrus, W.10). Today California has huge levels of resource consumption and wealth accumulation, which, of course, have social and environmental costs. However, the region is also the birthplace of innovative new technology and is home to some of today's most environmentally conscious people, searching for solutions to inherited problems.

Introduction to Crops and Cultivation

W.05

These sunny regions have sprouted major industries to grow our supermarket salads, fruits and other crops year-round. Huge areas of tomatoes in California, and peppers under plastic in Greece and Spain, need food, water and protection against pests and diseases. Pressure is mounting to reduce subsidies on water and to move to low-input, energy-efficient, self-sustaining and diversified farming. Just the kind of crops that a Greek or Spanish grandparent would have used – the ones that made the ancient Mediterranean healthy, wealthy and wise. Back for a future?

Fruits of the Mediterranean   W.06

Loquats and kiwis (Chinese gooseberries) from China, apricots from Iran, and sweet almonds from the Middle East have moved to the Mediterranean and California to soak up the sun and the water – from the irrigation lines. Mediterranean almonds are now one of the most important nuts in world trade.

Cork   W.07

When *Quercus suber* is 25–30 years old it will do its first strip – for cork. Around 15 billion wine corks are pulled a year, and trees produce around 4,000 corks per strip. Tree destroyer? Definitely not. Cork oaks, unlike most trees, regenerate their bark. This can be (skilfully) stripped off at around nine- to twelve-year intervals for up to 200 years without harming the tree at all.

Good for rural employment: Cork oak wood pastures, known as *montados* (Portugal) or *dehesas* (Spain), can also 'grow' charcoal and meat. Iberian pigs producing high-value ham, *jamon serrano*, feed on the fallen acorns.

Good for the environment: These managed wood pastures provide valuable habitats for many plants and animals including the Iberian lynx and 42 species of birds, among them the rare black vulture and the short–toed eagle. So buying real cork conserves the cork trees and the habitat and supports livelihoods.

Supporters: The **RSPB** works to protect the cork oak woods, and promotes industries that benefit local rural communities. They are the UK partner of BirdLife International, the global alliance of bird conservation organizations. www.rspb.org.uk

APCOR, the Portuguese Cork Association, campaigns to promote cork and the Portuguese cork industry, nationally and worldwide. www. apcor.pt

Art: Heather Jansch used cork and deadwood to sculpt pigs, piglets and the white stork of the open woodlands after visiting the cork oak forests. Heather lives and works in Devon.

Peppers W.08

Capsicums, peppers and chillies, which come from the New World, contain a hot compound, capsaicin, in the fruit's inner wall. Measured in Scoville units, mild chillies come in at around 600 units, but the really hot ones get up to 350,000 units. As well as spicing up our food, chilli extracts have been added to sprays, paints and rubber coatings to ward off insects, elephants, aquatic molluscs and rats.

Vegetables and Herbs W.09

Spice up your salad with the colourful, slightly bitter radicchio and peppery rocket. Garlic, coriander and basil give dishes a Mediterranean flavour. It is possible that basil only reached the shores of the Mediterranean in the 16th century – it originally comes from India and the Middle East.

Citrus W.10

This promiscuous family is fond of breeding. The clementine is a cross between mandarins and bitter Seville oranges, and the tangelo the offspring of tangerines and grapefruits. Citrus are used in perfumes, cleaning products, animal feed, anti-bacterial agents and as CFC substitutes, as well as for food and drink. In Florida citrus plantations need around 127 cm of water a year; twice as much as the annual rainfall. This means irrigation. As sources of river water, such as the Rio Grande, diminish, bore holes have been used. The Californians are now turning to solar-powered pumps to pump the water out of the ground at over 1,350 litres per minute. Environmentally friendly? If watering isn't carefully controlled, salts can build up in the soil, reducing yields. Today farmers are using computers to monitor systems and keep irrigation and fertilization to a minimum, conserving resources and reducing soil salinization.

Grape Vines   W.11

Dionysus, 'most gentle and most terrible'

The vine has sometimes been linked to immortality. Why? Because its short-lived fruits are given long life as wine, because the dead-looking vines burst into life each spring, because we have kept the vine alive through myth and stories for centuries, or maybe because of the effect wine has on us. Certainly mortals were taken to new heights when they drank wine as an offering to the gods in classical times.

*Art: **Tim Shaw**, Cornwall-based sculptor, is creating a wild Bacchanal – eventually some 20 dancing maenads will mirror the twisting shapes of the vines. www.timshawsculptor.com*

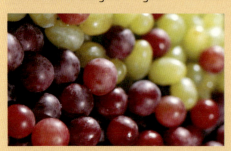

Dionysus, a potent nature god, here depicted as a bull, was associated with wine, fertility, festivities, intoxication, illusion and destruction. Also known as Bacchus, he started out with good intentions as the god of vegetation. However, things changed when he went from growing the vine to drinking its fermented juices … party time!

The land, like Dionysus, has changed. Here he stands, straddled between the ancient cultivated landscapes of the Mediterranean and the irrigated lands of today's intensive agriculture. For better or for worse?

Water solutions

In some Australian vineyards they practise 'partial root drying' where roots on one side of the vine are kept dry while roots on the other side are irrigated. The process is then reversed. The result: fewer side shoots, less pruning and more grapes.

Pulses W.12

Pulses, edible protein-rich seeds of the legume (pea and bean) family, can be dried, stored and easily transported. For some: a tasty treat. For some: a means of survival.

What can make food out of fresh air?

We need nitrogen to make protein. The air is 78% nitrogen, but animals and most plants can't absorb it like this. Peas and beans can. A type of bacteria, called a rhizobium, lives in their roots. In exchange for a home and sugar from the plant the rhizobia turn the nitrogen from the air into plant food. This improves people's diets, improves nutrient-poor soils, reduces inputs of artificial fertilizers and improves farmers' incomes.

Tasty food, toxic food?

As well as 'essential' amino acids, used to make our proteins, some legumes contain 'anti-nutritional compounds'. Some of these cause indigestion, others can kill.

Partner: Future Harvest International Centre for Agricultural Research in Dry Areas. *In times of crop failure or drought in Ethiopia, India and Pakistan people often eat grass peas, the only food around. These peas are harmless in small quantities, but over a long period can cause permanent paralysis. The alternative is starvation. Future Harvest have now bred a grass pea with low toxin levels that can be eaten without fear of paralysis.*

Tobacco W.13

Brought to Europe by Columbus, tobacco was heralded as a miracle medicine by many, and as an agent of the devil by a few, centuries before we realized that it was linked to cancer. The tobacco trade was associated with high profits, much of which were linked to government tax revenues, greed, piracy, smuggling and the slave trade.

Tobacco nicotine is more addictive than alcohol, heroin or cocaine. It has been estimated that by 2025 tobacco will probably kill as many people worldwide as dysentery, pneumonia, malaria and TB combined. 80% of tobacco is grown in the developing world. Tobacco depletes soil fertility, requires many sprays and is a cause of deforestation, but it does provide a means of livelihood for many people.

Tomatoes   W.14

The Conquistadors brought tomatoes from South America to Spain in the 16th century. At first the fruits were thought to be poisonous, but the Italians started to eat them after an intrepid herbalist pronounced them edible in 1544. Their reputation as an aphrodisiac comes from a French mistranslation of *pommi de mori* (apple of the Moors) as *pomme d'amour* (apple of love). Nowadays this ubiquitous culinary ingredient is more useful as a source of vitamins A, C and E.

Supporters: The Tomato Growers Association and **Delfland Nurseries** are helping with our tomato exhibit. www.britishtomatoes.co.uk

Grains   W.15

Grains feed the world because although they are small they are carbohydrate-rich and, like pulses, can be dried, stored and easily transported. The major grains are wheat (see O.17), rice (see H.13), maize and sorghum. Less familiar tef and millet are vital to the diets of local people.

Zea mays – *Zea* means the 'cause of life' and *mays* 'our mother'.

From New World ...

Maize was first farmed about 6,500 years ago in southern Mexico. It was deemed sacred, and Xilonen, the goddess of young maize, was sacrificed in Aztec times in its honour. Native Americans grew it with beans and squash, and together the crops were called 'the three sisters'. Niacin (vitamin B4) is difficult to absorb from maize. Lack of it causes pellagra, a serious disease that brings diarrhoea, dermatitis, dementia and death. Indigenous people prepared maize with alkaline ashes, which released the niacin, and also got the vitamin from their beans.

Partner: Future Harvest Maize and Wheat Improvement Centre. *Maize feeds millions in Africa, Asia, and Latin America. Unfortunately it has very low levels of lysine and tryptophan – essential amino acids. Scientists at the Future Harvest Maize and Wheat Improvement Centre in Mexico gained the World Food Prize in 2000 for developing, by conventional breeding methods, quality protein maize (QPM) with around double the usual level of lysine and tryptophan. In Ghana it is known as Obatampa (good nursing mother) and is fed to weaning babies without fear of nutrient deficiency. Two-thirds of all the maize grown is used as animal fodder. In China QPM maize, used as pig fodder, has provided a road out of poverty for many families.*

Mexico has a huge diversity of maize in a variety of shapes, colours and sizes. This diverse genetic mix helps to ensure healthy crops that are stress-resistant.

... to Old World

When maize was adopted in Europe and Africa, the beans and ashes were left behind. Pellagra became a common disease of poverty and slavery where diets were maize-dependent.

In the UK maize provides not only cereals and popcorn but also starches sheets, sweetens toothpaste, strengthens magazine papers and also crops up in glue, crayons, cosmetics, fireworks, inks, marshmallows, soaps, shoe polish – and more recently in fuels, plastics and biopharmaceuticals.

Sorghum bicolor: From Africa, this drought-resistant crop is the fifth most important cereal in the world. White-seeded forms make bread while astringent red-seeded types make beer.

Tef: Possibly the smallest-seeded grain crop in the world, tef provides a quarter of Ethiopia's cereal and is used to make injera, a fermented pancake-like bread.

Pearl millet: An important African cereal, adapted to hot, dry conditions, which grows on soils too poor for sorghum and maize.

Finger millet: Grows in the more favourable conditions. Its tasty seeds, rich in methionine (lacking in the diets of millions), can be stored for years.

Cotton   W.16

Cotton, the world's biggest non-food crop, makes half of the world's textiles. The long seed fibres are used for textiles, banknotes, fishing-nets, nappies, wallpaper and much more. The short fibres provide cellulose used for dynamite, sausage skins, photographic film, moulded plastic, and to thicken ice cream. The seeds themselves are pressed for oil and the seed meal is used for cattle feed, fish bait and fertilizer.

The cotton trade was a driving force in the Industrial Revolution and helped to finance the British Empire. The taste for Indian cotton calicoes helped to start the British cotton industry. Spinning machines brought poverty to thousands of villages in the Midlands when spinsters were no longer needed. Then imported coloured calicoes were outlawed, so that the water- and then steam-powered Lancashire spinning and weaving mills would survive. The giant mills undercut and ruined the village hand-craft cloth producers of India. The cotton for Manchester was grown and picked by African slaves in the American southern states.

Today, grown in America, Central Asia and sub-Saharan Africa, this natural fibre has survived competition from synthetics. It can also be an environmental villain, using a lot of fertilizers, pesticides and water. However, cotton provides a source of income for many people. In order to reduce pesticide use some are turning to genetically modified cotton, some to integrated pest management (which reduces chemical inputs) and others to organic production, currently less than 1% of the market but on the up.

Olives W.17

Several hundred years ago olive oil provided light for lamps, and the golden essence to anoint the brave, wise and rich and embalm the dead. Though it is thought to reduce cholesterol levels and deter heart disease, olive oil's uses are not just culinary; it also makes a good hair tonic.

Production is booming, with Spain taking the lead, and the squeeze is on to reduce chemical inputs to keep the land in good heart as well as the people.

Art: Debbie Prosser is a Cornish potter: 'I've made raw fired decorated domestic ware for twenty-one years. In Germany I studied Iron Age pots, a perfect springboard to invent a clay body from Cornish raw materials to make the olive oil vats for Eden.'

Plants as Perfume

W.18

The scent of violets, a whiff of mint – how do they make you feel? Scent goes straight to the seat of emotion and memory in the ancestral core of your brain. Plants use scent to attract pollinators and repel predators. Perfumiers make scents from plant extracts just as musicians use notes to compose melodies. Why do we use perfume? To signal, seduce or warn, like plants, or for sweet memory and comfort? Cleopatra, queen of perfume, power and seduction, wore kyphi (containing rose, crocus and violet) on her hands and aegyptium (almond oil, honey, cinnamon, orange and henna) on her feet. She also scented the purple sails of her barge.

Our five senses, smell, touch, sight, taste and hearing, give us information about our environment. Chemoreception, the detection of chemical signals in the environment, was probably the first sense to appear in primitive organisms, and developed into smell and taste. Humans can now detect over 10,000 different odours.

'When we give perfume to someone we give them liquid memory.' **Diane Ackerman**

'Smell is a potent wizard that transports us across thousands of miles and all the years we have lived... odours instantaneous and fleeting, cause my heart to dilate joyously or contract with remembered grief.' **Helen Keller**

Aromas and perfumed gardens are sometimes used in reminiscence therapy to help people with Alzheimer's. Geranium oil from *Pelargonium graveolens*, left, is an important and valuable oil in perfumery.

Supporters: The Body Shop Foundation and **QUEST** have helped to develop our perfume exhibit. www.the-body-shop.com, www.questintl.com

Art: Bill Mitchell and **Dave Mynne** join forces once again (see the Spice Boat, H.23) to bring you the Perfume Cart. This time let your nose do the walking as the cart takes you through the smells of the world and into the depths of your memory.

Imagine...
Building a World from Nothing

The Big Build

Facts from the Pit

The brief: create a spectacular theatre in which to tell the story of human dependence on plants.

The team found a disused china clay pit, over 60 metres deep and the size of 35 football pitches, which was sheltered, south-facing and spectacular – 'an architect would fall over backwards wanting to build something in it,' said David Kirkland of Nicholas Grimshaw & Partners. Jerry O'Leary, Works Manager, called it the biggest sand pit in the world. The bad news: it was an inverted cone shape with little level ground, unstable, prone to flooding, and contained no soil.

The level of the bottom was raised 17 to 20 m by slicing off the tops of the spoil heaps surrounding the pit. Twelve dumper trucks and eight bulldozers shifted 1.8 million tonnes of dirt in six months. Then, near-disaster: 43 million gallons of water (approx. 195 m litres) rained into the pit in three months. Our engineers came up with the drainage system to end all drainage systems. It can easily handle the 22 litres/sec of water that runs into the pit (that's 20,000 bathfuls a day).

Dodgy slopes were shaved back to a safe angle and terraces chopped out. Two thousand rock anchors, some up to 11 m long, were driven into the pit sides to stabilize them, and a 'soup' of plant seed and plant food sprayed on the slopes to knit the surface together.

Facts from the Biomes

In front of you are the biggest conservatories in the world, beautiful, iconic and super-efficient. The Humid Tropics Biome, which could hold the Tower of London, is 11 double-decker buses high and 24 long, with no internal supports. We made the *Guinness Book of Records* for the largest free-standing scaffold ever built; 12 levels, 25 m across, with 46,000 poles which laid end to end would stretch 230 miles.

Building a lean-to greenhouse on an uneven surface that changed shape was tricky. Bubbles were used as they can settle perfectly on to any shaped surface. The bubbles were made of hexagons, copying insects' eyes and honeycombs – commonsense nature; producing maximum effect with minimum resources. The Biomes' steelwork weighs only slightly more than the air they contain. They are more likely to blow away than blow down, so they are anchored into the foundations with steel ground anchors: 12-metre tent pegs! The final design comprised a two-layer steel curved space frame, the hex-tri-hex, with an outer layer of hexagons (the largest 11 m across), plus the occasional pentagon, and an inner layer of hexagons and triangles (resembling huge stars) all bolted together like a giant Meccano kit. Each component was individually numbered, fitting into its own spot in the structure and nowhere else.

The transparent foil 'windows', made of 3 layers of ETFE (ethylenetetrafluoroethylene-copolymer), form inflated 2-metre-deep pillows. ETFE has a lifespan of over 25 years, transmits UV light, is non-stick, self-cleaning and weighs less than 1% of the equivalent area of glass. It's also tough: a hexagon can take the weight of a rugby team. The pillows were installed by 22 professional abseilers – the sky monkeys.

Making the Earth

We needed topsoil, but it was simply not available in the quantity and quality we required without bringing it long distances. So we made our own; 85,000 tonnes of it. Soil is made of minerals of different sizes (sands and clays) and organic matter, mixed together in the right proportions. Soil manufacture has been tried before, but nowhere with so much at stake. The team worked within a tight budget and time-frame using wastes and recycled materials where possible, and avoiding peat.

> ***Science:*** *Dr. Stephen Nortcliff of the* **University of Reading**, *www.rdg.ac.uk, helped to develop Eden's soils, and the worms from* **Wiggly Wigglers** *dug and fertilized the new earth.* www.wigglywigglers.co.uk

Local mine wastes provided the minerals; the china clay company IMERYS had a bit of sand spare, and WBB Devon Clays Ltd had some reject clays.

The organic matter needed to be tough and long-lived, especially in the Biomes, where composted bark from the forestry industry was used. Plants in the Humid Tropics grow rapidly all year round and needed a high-performance soil capable of holding lots of water and nutrients. In the Warm Temperate we want to show how plants function in their natural, dry, harsh environment, so we put more sand in the soil so that it holds less water and nutrients. Bark and clay made up the rest of the

mix. Fynbos plants from South Africa don't grow well in fertile soil – for them it is toxic. This mix therefore had to be almost nutrient-free, and so consists simply of composted bark and sand. Outdoors, where the climate is less demanding, we went for composted domestic green wastes. The ingredients were mixed together in a nearby clay pit, like making a giant cake with a JCB.

We might have had faster growth using high-quality agricultural soil, but it wouldn't have been as significant. Eden shows that environmental regeneration is possible. Our soils technology has an application in the wider environment that we hope will have a massive impact in the future.

The Climate

Water: At the bottom of the pit Eden is 15 m below the water table, so without a state-of-the-art drainage and pumping system, it would be the Atlantis Project. The 'grey' water is collected and used for irrigation and flushing the loos. The rain falling on the Biomes is collected, treated and then used for irrigation and humidification within the greenhouses.

Eden's artificial climates are constantly monitored and controlled automatically.

Humidity: In the Humid Tropics we keep it up with automated misting nozzles (90% at night, and down to 60% during visiting hours). In the Warm Temperate we keep it down; vents are often open, even during relatively cool periods, to reduce humidity close to the leaves (which can otherwise cause fungal problems).

Temperature: The main heating source for both Biomes is the sun. The back wall acts as a heat bank, releasing warmth at night. The two layers of air in the triple-glazed windows give maximum insulation. Extra heating is provided through the air-handling units, the big grey boxes outside the Biomes. The Humid Tropics Biome ranges from 18°C to 35°C; and the Warm Temperate averages 25°C in summer with a winter minimum of about 10°C.

Ventilation: The vents may seem small for a building this large. They work because the height generates a 'chimney effect' that draws air through the system. On very hot days the air-handling units help circulate the air within the Biomes.

> **Supporter: Pennon Group Plc** (and their subsidiary, **South West Water Limited**) have been a great help to the project as a whole and have enabled us to have our fantastic waterfall in the Humid Tropics Biome.
> www.pennon-group.co.uk   www.swwater.co.uk

The Life

The plants

Eden is not a collection of rare plants but of the common plants of the world, the ones we depend on every day but many of us have never even seen.

Where did our plants come from? Most were already in cultivation in Europe and came to us from other botanic gardens, research collections, private individuals or from commercial nurseries. Very few were collected from the wild. Many plants arrived as seeds or cuttings and were, and still are, grown on at Eden's nursery. The rare plants we do have are there to tell a story and have been gathered with the full support of governments, and conservation and development organizations.

A book on the plants of the Humid Tropics is available, and others will follow.

Supporter: The Royal Botanic Gardens, Kew, have provided invaluable help with horticultural science in a wide range of ways, www.rbgkew.org.uk, and others such as **International Institute for Environment and Development** have contributed to our understanding of how lives depend on plants. www.iied.org.

The animals

We want to show biodiversity, but we don't have facilities for animals. So we portray quite a few, such as the cork pigs, through the work of artists.

We only need to pollinate the flowers if we want them to produce seeds or fruits. Many of our plants are wind-pollinated and many of the insect-pollinated ones accept the services of any passing insect. For special cases, however, we use a member of staff with a paintbrush!

We do get a few unwanted animals – otherwise known as pests – and run a rigorous healthcare programme. Isolation houses at Eden's nursery catch problems before they reach the pit. On site we use an integrated pest-management system with insect predators (biological controls) and 'soft' chemicals (soaps and oils) as required. Strings you might see dangling from the taller trees are small pulley systems to insert beneficial insects into the canopy. In some areas lightboxes emit UV light at night to catch moths and mosquitoes.

We have introduced several larger animals into the Biomes, most of which came from Newquay Zoo and have been bred in captivity. They help keep the plants healthy; birds, lizards, treefrogs and praying mantids consume pests in the canopy and problematic insects on the ground, day and night, giving us a 24-hour-a-day pest control.

Staff care for the animals and regularly monitor where they are. We would be grateful if you could report any sightings to the staff (except for the birds, which are not hard to spot). In the Humid Tropics Biome you may see: brown and green Anole lizards, also known as American chameleons, because they can change colour to match their surroundings; tropical house geckos; praying mantids (one is called Priscilla); white treefrogs nicknamed dumpy frogs due to their large size; greeny-yellow insectivorous birds called Sulawesi white eyes; and red jungle fowl that patrol the Biome controlling the pests in the plants and on the ground.

In the WTB can be found Moorish geckos (they look like miniature crocodiles) and meadow lizards that bask on the warm rocks. The robins are here on holiday! To ensure the health and safety of our animals please do not feed or handle them.

How Green is Your Valley?

Every individual, business and organisation has an environmental and social impact. Sustainability is about minimizing the impact of our actions, and the debate about sustainability is happening on all levels worldwide. As we think about our impact we ask you to think about your world, and your impact. Become involved!

At Eden we are tackling the difficult issues of waste, water, energy and the day-to-day impact of the whole Project. In fact we want to do more than just solve our problems, we want to solve them in an open way that makes our buildings, our practices and our organization part of the education. For many of our challenges there are no easy answers, and every issue requires vigorous debate. That isn't an excuse, but it is an important process that will enable us to make the right long-term decisions, which need to be balanced between environmental, social and financial aims.

We already have some major achievements to be proud of. Our Biomes are among the most efficient structures ever seen, using minimal materials that are long-lived and easily recyclable. We source the vast majority of our supplies locally, reducing transport energy and supporting local economies; our soils reuse waste; our water is recycled groundwater and harvested rainwater. We buy our Green Tariff electricity from Cornish wind farms, and run many of our vehicles on LPG. We are implementing a green travel plan for staff and visitors, an important element of the government's integrated transport policy aimed at promoting greener, cleaner travel choices.

We aim for efficiency wherever possible – using less is one of the greatest steps towards sustainability. Our giant greenhouses are working better than we ever anticipated, using only a third of the engineers' estimates for energy to heat them.

So what's next?

New Works

The dream was always to have three covered Biomes and a revolutionary Education Centre at Eden. The dreaming is over and we are now gearing up to complete Eden's tapestry.

The Dry Tropics Biome

Located above the Eden Village, this Biome will speak up for those people who live on the edge – in the hard, waterless parts of the world, where every day could bring famine, disease or resource wars. The Dry Tropics are as culturally rich as they are resource poor, occupy 1/3 of our world and are home to 1/4 of the earth's population.

The Education Centre

Thank you for bearing with us while we have piloted ideas and programmes in our temporary tented home. We look forward to inviting you to our new Education Centre – an inspirational hub for events, exhibitions and learning and an exhibit in its own right, inspired by plant form and crafted from plant materials.

Assisted Routes

We know it's a long walk round, we know some parts of the site are steep and so we are working on ways to make it easier.

The New Works mean some things on the site may move around a bit, and we apologise in advance for any inconvenience. As in the first 'Big Build' we hope to take you on the journey with us.

Thank you ...

We would like to say a big thank you to all of you who have supported us over the years. The list grows so rapidly that it is sadly impossible to mention everyone here. We are listing our supporters on the website, **www.edenproject.com**. You know who you are, so if you are not on the list, and should be, please let us know.

Last but not least thanks again to the Eden Team and the entire construction, design and professional team who have given beyond the call of duty and whose talents are imprinted right across the site.